97-REM-2

PATHOGEN DESTRUCTION EFFICIENCY IN HIGH-TEMPERATURE DIGESTION

by:

Donald M.D. Gray (Gabb)
East Bay Municipal Utility District

John M. Hake
East Bay Municipal Utility District

2004

WERF

The Water Environment Research Foundation, a not-for-profit organization, funds and manages water quality research for its subscribers through a diverse public-private partnership between municipal utilities, corporations, academia, industry, and the federal government. WERF subscribers include municipal and regional water and wastewater utilities, industrial corporations, environmental engineering firms, and others that share a commitment to cost-effective water quality solutions. WERF is dedicated to advancing science and technology addressing water quality issues as they impact water resources, the atmosphere, the lands, and quality of life.

For more information, contact:
Water Environment Research Foundation
635 Slaters Lane, Suite 300
Alexandria, VA 22314-1177
Tel: (703) 684-2470
Fax: (703) 299-0742
www.werf.org
werf@werf.org

This report was co-published by the following organizations. For nonsubscriber sales information, contact:

Water Environment Federation
601 Wythe Street
Alexandria, VA 22314-1994
Tel: (800) 666-0206
Tel: (703) 684-2452
Fax: (703) 684-2492
www.wef.org
pubs@wef.org

IWA Publishing
Alliance House, 12 Caxton Street
London SW1H 0QS, United Kingdom
Tel: +44 (0) 20 7654 5500
Fax: +44 (0) 20 7654 5555
www.iwapublishing.com
publications@iwap.co.uk

Library of Congress Catalog Card Number: 2003113629

IWAP ISBN: 1-84339-696-3 WEF ISBN: 1-57278-194-7

East Bay Municipal Utility District

The research on which this report is based was funded, in part, by the United States Environmental Protection Agency through Cooperative Agreement No. CR-825237-01 with the Water Environment Research Foundation (WERF). Unless an EPA logo appears on the cover, this report is a publication of WERF, not EPA. Funds awarded under the Cooperative Agreement cited above were not used for editorial services, reproduction, printing, or distribution.

ACKNOWLEDGMENTS

The authors acknowledge the efforts of Carlos DeLéon who was primarily responsible for operating the Phases 1 and 2 anaerobic digesters. The authors extend additional thanks to: Alex Borys, Vince DeLange, Scott Nilson, Bob Poole, and Lisa Servande at the EBMUD Main Wastewater Treatment Plant; the EBMUD laboratory staff, especially Ken Gerstman and Jack Lim; the BioVir Laboratory, Inc. staff; and a cadre of Environmental Careers Organization (ECO) interns including Katie Alcantar, Carl Anderson, Daria Evans-Walker, Chanice Harris, David Heber, Chisato Nakade, Atthasit Pongsakdi, Philip Ting, and Xiaozhou You. Financial contributions to support the project were received from the Bay Area Clean Water Agencies (BACWA) and the Sacramento Regional County Sanitation District (SCRSD).

Report Preparation

Principal Investigator:
Donald M.D. Gray (Gabb), Ph.D., P.E.
East Bay Municipal Utility District

Project Team:
Richard E. Danielson, Ph.D.
BioVir Laboratories, Inc.

Carlos DeLéon
University of California, Berkeley

Sambhunath Ghosh, Ph.D.
Enviro-Energetics

John M. Hake, P.E.
East Bay Municipal Utility District

David Jenkins, Ph.D.
David Jenkins and Associates, Inc.

David R. Williams, P.E.
East Bay Municipal Utility District

Project Subcommittee

Cecil Lue-Hing, D.Sc., P.E.
Cecil Lue-Hing and Associates

Ronald B. Sieger, P.E., DEE.
CH2M Hill

James E. Smith, D.Sc.
U.S. Environmental Protection Agency

Water Environment Research Foundation Staff

Director of Research:	Daniel M. Woltering, Ph.D.
Project Managers:	Amit Pramanik, Ph.D.
	Roy Ramani, Ph.D.

ABSTRACT AND BENEFITS

Abstract:

The purpose of this research was to evaluate and compare various thermophilic anaerobic digestion processes for meeting U.S. EPA biosolids Class A pathogen standards. The project was split into three phases. Phase 1 screened three bench-scale thermophilic anaerobic process configurations at three different thermophilic temperatures based on their fecal coliform destruction efficiency. All three of the thermophilic process configurations tested were capable of achieving the Class A fecal coliform standard and were included in Phase 2. In Phase 2, bench-scale anaerobic digesters were fed primary sludge seeded with *Escherichia coli*, helminth ova, poliovirus, and *Salmonella* to evaluate pathogen destruction. Two process configurations, the thermophilic single-stage and the two-stage mesophilic acid-phase/thermophilic methane-phase system, met Class A requirements at 50°C. In Phase 3, the single-stage thermophilic anaerobic digestion process was compared to the single-stage mesophilic process at full scale (1.5-MG digesters) based on fecal coliform and pathogen destruction, process performance, digested sludge dewaterability, and odor generation.

Pathogen destruction and process performance comparisons of the various process configurations are presented for each phase of the study. Based on the fecal coliform data presented here, an empirical model was developed for quantitatively comparing multiple stage and single-stage thermophilic anaerobic digester performance. The model demonstrates that various combinations of thermophilic temperatures, staging, and residence times can achieve the Class A fecal coliform requirement. This study also suggests that anaerobic digesters operating in the lower thermophilic temperature range (approximately 50°C) are not only capable of achieving Class A requirements but may also produce digested sludges with less odor and lower volatile solids than digesters operating at higher thermophilic temperatures.

Benefits:

- Provides novel, patented (U.S. Patent No. 6,447,683 B1) thermophilic processes for meeting U.S. EPA Class A pathogen standards.
- Presents directly comparable data on pathogen destruction and process performance for several thermophilic anaerobic digestion process configurations.
- Demonstrates at bench scale that two thermophilic anaerobic digestion process configurations can achieve Class A requirements at 50°C, per EPA's Alternatives 3 and 4.
- Provides an empirical model for quantitatively evaluating the disinfection capabilities of different thermophilic anaerobic digestion process configurations based on fecal coliform destruction.
- Demonstrates at bench scale and in part at full-scale that thermophilic anaerobic digesters operating in the lower thermophilic temperature range (approximately 50°C) can not only achieve Alternatives 3 and 4 Class A pathogen requirements, but also may produce digested sludges with better volatile solids destruction and less odor than digesters operating at higher thermophilic temperatures.

- Describes methods for evaluating helminth ova destruction in a full-scale anaerobic digester when the feed sludge concentrations are too low to demonstrate a 2-log reduction.
- Demonstrates a possible relationship between thermophilic anaerobic digester temperature and propionic (and other volatile) acid concentrations in digested sludge. This finding may allow WWTPs to obtain Class A biosolids using thermophilic anaerobic digestion processes and minimize odor impacts.
- Demonstrates a possible inverse relationship between thermophilic anaerobic digester temperature and volatile solids reduction efficiency.
- Demonstrates that the capillary suction time (CST) test may not be a suitable indicator of dewaterability for comparing thermophilic and mesophilic digested sludges.

Keywords: thermophilic, anaerobic digestion, biosolids, Class A, digester

TABLE OF CONTENTS

LIST OF TABLES

LIST OF FIGURES

LIST OF ACRONYMS

°C	degrees Celsius
Cc	cake solids concentration
Coeff. of var.	coefficient of variation
Cr	centrate solids concentration
Cs	digested sludge solids concentration
CST	capillary suction time
CSTR	continuous stirred tank reactor
DCS	distributed control system
EBMUD	East Bay Municipal Utility District
°F	degrees Fahrenheit
g	gram
gpd	gallons per day
gpm	gallons per minute
h, hr	hour
HRT	hydraulic residence time
HX	heat exchanger
L	liters
m^3	cubic meters
MCRT	mean cell residence time
Meso	mesophilic
Meso 1	mesophilic control
Meso AP	mesophilic acid-phase
Meso MP	mesophilic methane phase
Meso/thermo	two-stage, mesophilic acid first stage followed by a thermophilic gas second stage
mg	milligram
MG	million gallons
MPN	most probable number
MWWTP	main wastewater treatment plant
PC	personal computer
PEC	Pathogen Equivalency Committee
PFRP	Process to Further Reduce Pathogens
PFU	plaque-forming units
PS	primary sludge
RTD	resistance temperature detector
s	second
SM 20	*Standard Methods for the Examination of Water and Wastewater,* 20th Ed.
st. dev.	standard deviation
thermo/meso	Two-stage, thermophilic acid first stage followed by a mesophilic methane second stage
TALK	total alkalinity
TPAD	temperature-phased anaerobic digestion
TS	total solids
TWAS	thickened waste activated sludge

Thermo	thermophilic
Thermo 1	thermophilic control
Thermo AP	thermophilic acid-phase
Thermo MP	thermophilic methane-phase
TVA	total volatile acid
U.S. EPA	United States Environmental Protection Agency
VAR	vector attraction reduction
VFA	volatile fatty acid
VOC	volatile organic compound
VS	volatile solids
VSR	volatile solids reduction
v/v	volume per volume
WWTP	wastewater treatment plant

EXECUTIVE SUMMARY

Introduction

Land application of biosolids that only meet U.S. Environmental Protection Agency (U.S. EPA) Class B pathogen standards (40 CFR 503) can be controversial because Class B biosolids can contain human pathogens. More and more counties are banning Class B biosolids land application. Wastewater treatment agency managers are concerned about the future of biosolids land application and are considering treatment upgrades to achieve U.S. EPA Class A standards, which require the virtual elimination of pathogens from biosolids. While meeting these standards would significantly broaden the range of opportunities for biosolids recycling, the processes required to achieve Class A pathogen levels are often cost-prohibitive. Low-cost processes for meeting Class A standards would provide attractive biosolids management alternatives. This research specifically examined thermophilic anaerobic digestion processes as alternatives to meet Class A standards.

At the beginning of the study, an extensive literature review was conducted to assess state-of-the-art thermophilic anaerobic digestion methods. The more recent of these studies described improvements that enhance stability and performance of the process. These improvements were either incorporated into the experimental design or evaluated as part of this research and include:

- Fast start-up method. The fast start-up method produces a more robust thermophilic culture that is tolerant of temperature fluctuations and increases process stability.
- Draw/fill operation. The draw/fill cycle improves pathogen removal by reducing short-circuiting.
- Multiple-stage digestion. Multiple-stage reactors reduce short-circuiting and, if the second-stage reactor is mesophilic, produce a less odorous sludge and supernatant. Phased systems may digest organic material more efficiently.

The objectives of the study were to:

1. Evaluate and compare candidate thermophilic anaerobic digestion processes for meeting Class A requirements;
2. Evaluate the thermophilic anaerobic digestion process at full-scale for stability, dewaterability, and odor generation; and
3. Apply for a Class A pathogen equivalency (PFRP) recommendation from U.S. EPA's Pathogen Equivalency Committee (PEC) for the thermophilic process selected for full-scale evaluation.

The effectiveness of thermophilic anaerobic digestion processes to achieve Class A was evaluated at bench-scale under laboratory conditions using U.S. EPA's Alternatives 3 and 4 (40 CFR 503 Section 503.32(a)(5) and (6), (demonstration of enteric virus and helminth ova reduction), rather than Alternative 1 (Section 503.32(a)(3), meeting time/temperature requirements).

The research was divided into three phases. In Phase 1, three thermophilic anaerobic digestion process configurations were screened at bench-scale and evaluated using fecal coliform destruction as a measure of disinfection efficiency. These processes were:

(1) thermophilic acid phase;

(2) one-stage thermophilic; and

(3) two-phase, mesophilic acid phase/thermophilic methane phase.

Each process configuration was examined at three thermophilic temperatures– approximately 48° C, 53° C, and 62° C, and compared with a mesophilic control at 35° C. Processes showing potential for meeting the Class A fecal coliform requirement (<1,000 MPN/g TS) were carried forward for further evaluation in Phase 2, where the feed sludge was seeded with pathogens and digester performance was evaluated on the basis of pathogen destruction. Those process/temperature combinations that met Class A requirements in Phase 2 were considered for evaluation in Phase 3, a full-scale pilot study. The objectives of the full-scale test were to confirm bench-scale pathogen destruction results and evaluate other aspects of digester performance including solids stabilization, process stability, dewaterability, and odor generation.

This process method was issued the United States Patent No. 6,447,683 B1 on September 10, 2002. The patent for "Pathogen Destruction Process for Biosolids" was assigned to the Water Environment Research Foundation, and it was dedicated to public benefit by WERF and the East Bay Municipal Utility District on November 21, 2002.

Phase 1 — Screening Evaluation

The purpose of Phase 1 was to evaluate several thermophilic anaerobic digestion processes and select one or more processes for Phase 2 pathogen testing. Selection was based primarily upon fecal coliform destruction although process stability was also considered. Three thermophilic anaerobic digestion process configurations were screened at bench-scale:

(1) two-stage thermophilic acid phase/mesophilic methane phase;

(2)one-stage thermophilic; and

(3) two-stage mesophilic acid phase/thermophilic methane phase.

Each process was examined at three thermophilic temperatures (about 48° C, 52° C, and 62° C) and compared against a mesophilic control at 35° C (referred to as Meso 35). Because the thermophilic acid stage in Process 1 was expected to provide a significantly greater disinfection effect than the mesophilic methane stage and to contain experimental costs, only the acid stage was operated in Phase 1.

Laboratory-scale Plexiglas digesters with an operating capacity of approximately 30 liters were used in Phase 1. The contents were mixed by an externally mounted variable-speed DC motor that turned a stainless steel shaft and two 4-in. diameter propellers. Each digester was heated by a 520-watt heat tape. Digester temperature was controlled by a proportional-response temperature controller that received a signal from an 18-in. submersible resistance temperature detector (RTD) probe.

Digester temperature appears to affect operating performance and stability. In the case of the acid-phase digesters (Process 1), the digesters with higher operating temperatures were characterized by higher total alkalinity concentrations and pH levels. Although propionic acid concentrations were very high in the acid-phase digesters (>1,200 mg/L), they were not substantially different among the digesters at different operating temperatures.

The same relationship does not appear to hold for the single-stage and methane-phase thermophilic digesters, where higher operating temperatures appear to be associated with higher total volatile acid (TVA) concentrations. For the single-stage thermophilic digesters (Process 2), alkalinities were not substantially different among the digesters, but the volatile acid concentrations, especially that of propionic acid (about 700 mg/L), were higher at higher temperatures. The thermophilic methane-phase digesters (Process 3) showed a similar result although alkalinities were lower at higher temperatures among these digesters.

The primary purpose of the Phase 1 screening evaluation was to determine which process configuration/temperature combinations could achieve the Class A fecal coliform criterion (<1,000 MPN/g TS). Graphical and statistical analyses were performed to evaluate the results from this phase of the study. There is a consistent inverse relationship between effluent fecal coliform density and digester operating temperature for each process tested (Table ES-1).

Table ES-1. Phase 1: Mean Log Fecal Coliform Concentrations.

Mean Operating Temperature (°C)	Process 1 Acid phase	Process 2 One-stage	Process 3 Two-stage
35	3.84	2.95	2.42
48	3.34	NA	0.81
49	NA	1.80	NA
52	NA	NA	0.37
53	NA	1.20	NA
54	2.01	NA	NA
62	1.48	0.72	0.34

The mean effluent fecal coliform concentrations were plotted against mean operating temperatures for each process tested in Phase 1. The means are compared to the Class A criterion of 1,000 MPN/g TS. Results for all three processes are plotted in Figure ES-1. The highest fecal coliform reductions were achieved by Process 3, the next highest by Process 2, and the lowest by Process 1. Because only the thermophilic acid-phase reactor was used to test Process 1 in this phase of the study, it was anticipated that actual Process 1 fecal coliform reductions would be higher than those shown in the figure.

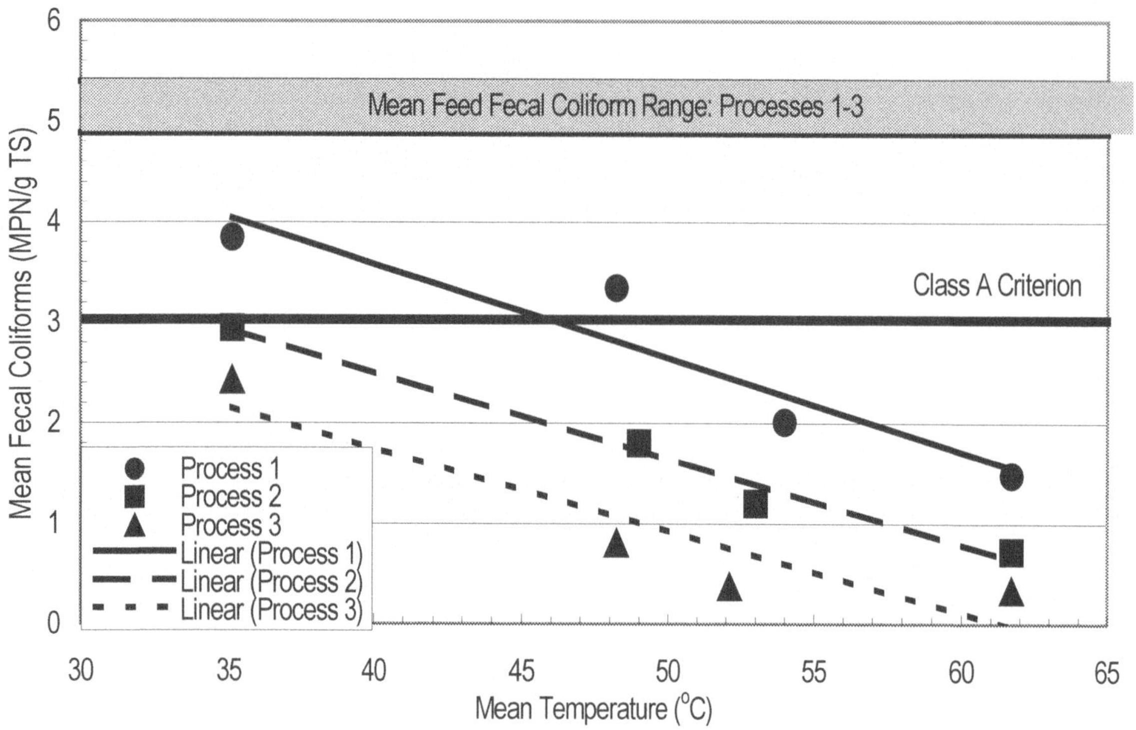

Figure ES-1. Phase 1 Mean Effluent Fecal Coliform Concentrations.

All three process configurations met the Class A fecal coliform requirement (Table ES-2) at the fecal coliform feed densities occurring during this phase of the project (fecal density means: 160,000–300,000 MPN/g TS, standard deviations: 170,000–860,000 MPN/g TS)

Table ES-2. Phase 1 Results for Meeting Class A Fecal Coliform Requirement.

Process	Temperature: Fecal Coliform Class A requirement met?			
	35	48-49	52-54	62
Process 1 Thermo Acid phase	No	No	Yes	Yes
Process 2 One-stage	No	Yes	Yes	Yes
Process 3 Meso Acid/Thermo Methane	Yes	Yes	Yes	Yes

Based on these results, all three processes were carried forward for further evaluation in Phase 2. The thermophilic digesters were to be operated at a single temperature, 50°C, because all three process configurations appeared to meet the Class A fecal coliform requirement at approximately 50°C. Further, thermophilic anaerobic digestion appeared to be more stable and in general less expensive to implement and operate at full scale if operated at lower thermophilic temperatures (≤ 50°C).

Phase 2 — Pathogen Destruction

The purpose of Phase 2 was to feed each process a municipal sludge seeded with pathogens and determine pathogen destruction capability. The seeded pathogens were *Salmonella*, poliovirus, and helminth ova. *Escherichia coli* were also added to increase the low background levels of fecal coliforms (approximately 10^5) in the EBMUD digester feed sludge.

The following four digester systems were operated:

1. Mesophilic single-stage (Meso 1 or Control) at 35°C;
2. Thermophilic single-stage (Thermo 1) at 50°C;
3. Two-phase, thermophilic acid phase (Thermo AP) at 50°C and mesophilic methane phase (Meso MP) at 35°C; and
4. Two-phase, mesophilic acid phase (Meso AP) at 35°C and thermophilic methane phase (Thermo MP) at 50°C.

Hydraulic residence times (HRTs) for the acid-phase, methane-phase, and single-stage digesters were 2, 13, and 15 days, respectively. Total HRTs were 15 days for each process configuration.

Erlenmeyer flasks (2 L) with an operating capacity of about 1.5 L were used as digesters. Their contents were mixed by magnetic mixers with floating stir bars. The digesters were heated by immersing them in constant temperature water baths equipped with thermostatically-controlled immersion heaters. Digester temperatures were continuously monitored and recorded by Type T thermocouples linked to a personal computer with data logging software. Peristaltic pumps were used for feeding and withdrawing sludge from the reactor. HRTs were controlled by adjusting flow rates and pumping intervals. Feeding was semi-continuous on a fill/hold/draw cycle. Digester feed was primary sludge (PS) from the East Bay Municipal Utility District (EBMUD) main wastewater treatment plant. Thickened waste activated sludge (TWAS) was not used for feed in Phase 2 to avoid nocardioform foaming in the digesters, which would be introduced with EBMUD TWAS.

The three thermophilic processes (Thermo 1, Thermo AP/Meso MP, and Meso AP/Thermo MP) consistently met the Class A criterion for fecal coliforms and substantially reduced unusually high *Salmonella* concentrations (4–7 log units reduced), while the control digester (Meso 1) did not. Of the three thermophilic processes studied, only the Meso AP/Thermo MP process consistently reduced enteric virus and helminth ova concentrations to the non-detectable levels required of Class A sludge (with one exception). After Day 14, Thermo 1 also consistently met virus and helminth Class A requirements (with one exception). Pathogen destruction did not appear to be affected by high-acid conditions in the digesters.

Thermo MP and Thermo 1 digesters most consistently achieved Class A requirements on the basis of fecal coliform, *Salmonella*, poliovirus, and helminth ova results. These digesters have the longest HRTs at thermophilic temperature (13 and 15 days, respectively). The Thermo MP, however, appears to perform slightly better than the Thermo 1 digester, even though the Thermo MP has a slightly lower thermophilic HRT. This suggests that staging digesters enhances disinfection.

Phase 3 — Full-Scale Pilot Test

Bench-scale testing conducted as part of the Screening Evaluation (Chapter 2.0) and Pathogen Destruction Test (Chapter 3.0) demonstrated that 2 thermophilic anaerobic digestion processes could achieve Class A standards for biosolids. The 2 processes, 2-phase Meso Acid/Thermo Methane and single-stage Thermo, were considered for full-scale testing. The single-stage Thermo was selected for full-scale testing primarily because it was easier to implement; i.e., full-scale single-stage digesters were available with a feed system that could be modified to the fill/hold/draw feed regime.

The full-scale pilot study was conducted primarily to confirm pathogen destruction results obtained at bench-scale, i.e., that a single-stage thermophilic anaerobic digester could achieve Class A pathogen standards for biosolids. In addition, process stability, dewaterability, and odor generation could be evaluated and compared to the performance of a single-stage mesophilic anaerobic control digester.

Two full-scale anaerobic digesters at EBMUD's MWWTP were selected for the pilot study and are referred to as Thermo Test and Meso Control. Operating criteria for these two digesters are shown in Table ES-3.

Table ES-3. Preliminary Operating Criteria — Full-Scale Test.

Item	Thermo Test	Meso Control
Target Operating Temperature	45-50°C	35°C
Target Hydraulic Retention Time (HRT)	15 days	15 days
Feed Sludge	Primary/TWAS mix (approx. 50/50 v/v)	Primary/TWAS mix (approx. 50/50 v/v)
Feeding Regime	fill/hold/draw	continuous
Target Loading Rate Range	2-3 kg TS/m^3-day	2-3 kg TS/m^3-day
Digester Diameter	95 feet	95 feet
Digester Volume	1.5 MG	1.5 MG
Digester Heating	External spiral-type HX	External spiral-type HX
Digester Mixing	Compressed digester gas	Compressed digester gas
Digester Cover Type	Floating	Floating

A fast thermophilic anaerobic digester start-up (changing rapidly from mesophilic to thermophilic temperature) is important to ensure the development of a robust thermophilic culture. To accomplish this, the Thermo Test digester was first filled with water and heated. Only after the water reached 45°C was the digester seeded with mesophilic-digested sludge and slowly fed. This provided the rapid start-up and temperature change from mesophilic to thermophilic operation. Also important is the feeding regime. Thermo Test was fed intermittently using fill, hold, and draw phases. The hold phase ensures that all of the feed sludge entering the thermophilic digester is retained for a minimum period of time, with no opportunity for short-circuiting to digester effluent. The feed cycle durations and digester sludge fill levels were calculated to achieve the target HRT of 15 days.

Two significant operating changes were made during the pilot study. The first was to increase the Thermo Test digester's temperature. Testing demonstrated that complete helminth ova destruction was not possible within a 6-hour hold period at 45°C (Period 1), so the operating temperature was increased to 50°C (Period 2). The heat exchanger for the thermophilic digester was unable to deliver sufficient heat to maintain the higher temperature at the target feed rate and so the feed rate was reduced, increasing HRT from 15–35 days. At the lower feed rate the digester temperature was maintained in the range of 48°–50°C.

The primary purpose of the full-scale pilot study was to demonstrate that the single-stage thermophilic anaerobic digester is capable of producing Class A biosolids at full-scale. Complicating the full-scale study, the U.S. EPA Pathogen Equivalency Committee (PEC) requires a sufficient number of pathogens be present in the feed sludge to demonstrate adequate destruction. Most raw wastewater sludges in the U.S. do not contain pathogens at high enough densities for this type of demonstration. It is impractical to seed full-scale digesters with pathogens because of the large number of organisms required, public health concerns, and cost. Instead, helminth ova destruction was determined using ova contained in Sentinel Chambers. The sentinel chambers are cylinders capped on both ends with a screen mesh that retains helminth ova but allows for liquid exchange with the surrounding media. The chambers used for this test were constructed from 1-in. diameter acrylic tubing with 30-µm stainless steel mesh glued to both ends of a 0.25-in. long tube.

The chambers were loaded with EBMUD digester sludge and seeded with helminth (*Ascaris suum*) ova. The chambers were placed in three wire mesh baskets that were attached to a wire cable at approximately 5-, 10-, and 15-ft depths. At regular time intervals, the cable was pulled out of the digester for chamber retrieval. Duplicates were obtained at each sample depth and period. Control chambers showed survivals $\geq$ 94% in both test runs. Temperature sensors were also placed at each sample station along the cable to continuously monitor temperatures. The goal was to measure viable helminth ova densities at a specific chamber removal time at all depths. This removal time would determine the hold phase interval to ensure that all sludge entering the digester was held for this minimum time period without any possibility of short-circuiting.

Helminth ova survival was assessed at full scale using sentinel chambers at two different thermophilic temperatures. The sentinel chamber test conducted at 45°C showed minimal helminth ova destruction, even after 12 hours, where the lowest ova viability observed was 75% in the bottom chamber. A second test was conducted at 50°C, and in this test no viable helminth ova were observed in any of the non-control chambers except for one. This chamber, a bottom sample taken after 12 hours in the digester, had 10 viable ova out of 1,178 counted (0.85% viability). Because all of the 29 other chambers contained no viable ova (especially those retrieved preceding the 12-hr chamber at 4, 6, and 8 hours), finding viable helminth ova in the 12-hr chamber was considered an anomaly. It was concluded from these results that the Class A helminth ova criterion was achieved in a semi-continuously fed single-stage thermophilic digester at 50° C with a minimum 4-hr holding period.

The Thermo Test digester consistently met the Class A fecal coliform criterion (<1,000 MPN/g TS) and fecal coliform densities were often undetectable. The Meso Control digester rarely met the Class A fecal coliform criterion despite the relatively low fecal coliform densities in the feed sludge (approximately 10^6 MPN/g TS) during Period 2 of this phase.

Generally, pH and total alkalinity (TALK) concentrations were higher in the Thermo Test digester operated at 50°C (Period 2) than in the Meso Control. Specific volatile fatty acid (VFA) concentrations were also higher in the Thermo Test (Period 2) digested sludge compared to the Meso Control sludge, e.g., acetic acid (170 mg/L vs. 115 mg/L) and propionic acid (55 mg/L vs. 37 mg/L). The higher operating temperature appears to result in a higher propionic acid concentration, a potential odor source. These observations are consistent with bench-scale results. However, the TVA/TALK ratio for Thermo Test was usually less than 0.10, and the maximum value was about 0.20, indicating reasonable process stability. The TVA/TALK ratio for Meso Control was generally lower than that for Thermo Test. VS reduction (VSR) generally ranged between 50% and 65% for both digesters, with no significant difference between digesters. Both digesters showed methane gas compositions in a range of 50% to 65%, which is typical for anaerobic digesters treating municipal wastewater sludges.

Previous studies (Drury et al., 2001; Garber, 1954; Golueke, 1958; Lee et al., 2001; Rimkus et al., 1982) have indicated that thermophilically-digested sludges dewater better than mesophilically-digested sludge. Thermophilic and mesophilic sludges were segregated to compare sludge dewaterability using full-scale horizontal solid-bowl centrifuges. Concurrently, capillary suction time (CST) tests were conducted in the laboratory. The Thermo Test (50°C) sludge dewatered better than the Meso Control (35°C) sludge, as shown by the higher cake solids (average values of 24.4% and 20.9%, respectively) and greater solids capture efficiency (average values of 87.7% and 74.2%, respectively). Thermo Test sludge cake had fecal coliform densities almost 2 logs lower than Meso Control's sludge cake and nearly always lower than the Class A fecal coliform criterion. These full-scale dewatering results were contrary to those indicated by the capillary suction time (CST) test results. Although these results indicated that Meso Control's digested sludge should dewater more readily than Thermo Test's sludge, the field results indicated the contrary. The CST test may not be a suitable indicator of dewaterability for comparing mesophilic and thermophilic sludges.

Digester gas and digested sludge samples were collected from the Meso Control and the Thermo Test digesters to quantify the relative odor-formation potential between the two anaerobic digestion processes. Full-scale pilot test digester gas samples showed that VOC levels were higher in the Thermo Test (50°C) digester than in Meso Control (35°C). Acetic acid levels were higher in the Thermo Test digested sludge. Results for propionic acid were mixed with a greater number of measurable concentrations from Thermo Test but the highest single concentration was observed in Meso Control. Hydrogen sulfide and other reduced sulfur compound concentrations were similar in gas produced by both digesters.

Summary and Conclusions

A substantial body of fecal coliform data was collected during each study phase, allowing for a comprehensive comparative analysis. An empirical model was developed to quantitatively compare two-stage (thermo/meso and meso/thermo) and single-stage thermophilic anaerobic systems.

With the 655 bench-scale and full-scale data points collected in this study, a model that predicts digested sludge (biosolids) fecal coliform density using digester sludge temperature, thermophilic mean cell residence time, staging, and feed sludge fecal coliform density was developed using multiple regression analysis. This model is shown below:

$$\mathrm{Log}\,(FC_f) = 4.19 - 0.0516\,TEMP - 0.0924\,TMCRT - 0.906\,STAGED + 0.211\,\mathrm{Log}(FC_i)$$

Where:

FC_f = the digested sludge (biosolids) fecal coliform density (MPN/g TS)

$TEMP$ = the digester sludge temperature (°C)

TMCRT = the digester sludge thermophilic mean cell residence time (days), which includes the MCRT only in thermophilic digesters and excludes the MCRT in mesophilic digesters. [For example a single-stage mesophilic digester with a 20-day MCRT has a TMCRT = 0, and a two-stage thermo/meso system with a 2-day MCRT thermophilic first stage and a 13-day MCRT mesophilic second stage has a TMCRT = 2 days.]

STAGED = this value is either 0 or 1, when STAGED = 0 the digester is single stage, when STAGED = 1 the digester is two-stage, with the first stage operated at a 2-day MCRT

FC_i = digester feed sludge fecal coliform density (MPN/g TS)

Using the model derived from the regression analysis, performance curves were constructed to predict fecal coliform destruction for each of the thermophilic anaerobic digestion process configurations tested under various conditions. Assuming an initial fecal coliform density of $10^{8.5}$ MPN/g TS, Figure ES-2 predicts the resulting digested sludge (biosolids) fecal coliform density for anaerobic digester systems operated at thermophilic MCRTs from 2–15 days and at 45°C, 50°C, or 55°C. The figure also compares thermophilic systems with and without staging. Since all first stage digesters of the two-staged systems were operated with a 2-day MCRT, the staged systems indicated at a 2-day thermophilic MCRT (*x*-axis) in Figure ES-2 represent a thermophilic first stage (with a 2-day MCRT) and a mesophilic second stage (with a 13–15 day MCRT), and all other staged systems (thermophilic MCRT = 4–15 days) represent a mesophilic first stage (with a 2-day MCRT) and a thermophilic second stage at the MCRT indicated on the *x*-axis.

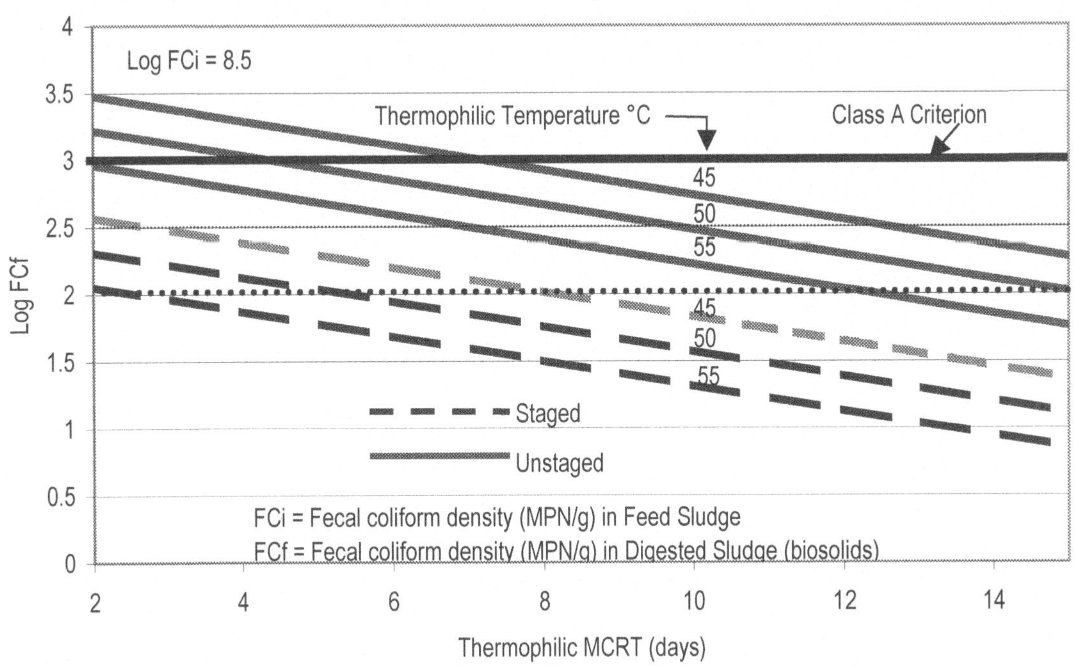

Figure ES-2. Performance Curves Based on Model-Predicted Fecal Coliform Destruction for Thermophilic Anaerobic Digesters Feed Sludge with a Fecal Coliform Density = $10^{8.5}$ MPN/g.

Principal Conclusions

This study's principal conclusions are:

- Staging anaerobic digesters improves fecal coliform destruction; however, single-stage thermophilic digesters operating in the fill/hold/draw sequence can outperform 2-stage digesters, by further reducing fecal coliforms, when single-stage digesters are operated at higher temperatures or longer thermophilic MCRTs or both. This relationship is quantified by a model presented in this report.
- Anaerobic digesters operated at lower thermophilic temperatures (45°C–50°C) may generate less odor than digesters operated at higher thermophilic temperatures (53°C–62°C) because less volatile acids (especially propionic acid) are formed at the lower thermophilic temperatures.
- Anaerobic digesters operated at lower thermophilic temperatures (45°C–50°C) may provide for higher VSR than digesters operated at the higher thermophilic temperatures (53°C–62°C).
- Anaerobic digesters operated at lower thermophilic temperatures can reliably meet U.S. EPA's Alternatives 3 and 4 Class A criteria if an adequate thermophilic MCRT is maintained. Because of its other potential benefits (e.g., less odor, higher VS reduction, and less energy consumed) operating anaerobic digesters in the lower thermophilic temperature range (45°C–50°C) may be preferable to operating at higher thermophilic temperatures (53°C–62°C).
- Depending on the thermophilic MCRT and temperature used, it appears that any of the thermophilic anaerobic digester configurations (i.e., single-staged thermophilic or two-staged — meso/thermo or thermo/meso) evaluated in this study can meet U.S. EPA's Alternatives 3 and 4 Class A pathogen requirements. Therefore, the digester configuration selected by a wastewater treatment agency can be based on ease of implementation rather than force-fitting a particular digester configuration into every wastewater treatment plant.

Additional conclusions are provided by chapter.

Phase 1 Screening Evaluation Conclusions

- At a 2-day MCRT, when digester thermophilic temperature rises:
  - TVA decreases
  - TALK increases
  - VSR increases
- At a 15-day MCRT, when digester thermophilic temperature rises, in general:
  - TVA increases
  - TALK decreases
  - VSR decreases
  - Specific VFA (such as propionic acid, which is linked to odor generation) increases
- Two-stage meso/meso (mesophilic acid first-stage followed by a mesophilic methane second-stage) meets Class A requirements for fecal coliform, but could not meet

Class A helminth ova requirements when ova-filled sentinel chambers were held in the process for 17 days.

Phase 2 Pathogen Destruction Conclusions

- Ammonia levels in digester sludge are significantly higher than that in feed sludge. This is probably due to deamination of organic nitrogen from the feed sludge in the digester.
- Digester sludge ammonia levels increase when the digester MCRT is increased from 2 to 15 days, and both systems are fed the same feed sludge.
- Digester sludge ammonia concentration increases with increased digester temperature (from 35°C to 50°C) at a 2-day MCRT. The digester sludge ammonia concentration appears to be independent of digester temperature at the longer MCRTs (i.e., 15 days and greater).
- Regressing Phase 2 data showed that the digested sludge (biosolids) fecal coliform density is significantly influenced by digester temperature, feed sludge fecal coliform density, thermophilic MCRT, and whether the digester is two- or single-staged.
- Regressing Phase 2 data showed that biosolids fecal coliform density is not significantly influenced by digester sludge TVA or TALK levels. Since anaerobic digestion process stability is usually determined by TVA/TALK ratio (< 0.10 is a stable process), this suggests that thermophilic anaerobic disinfection may not depend on process stability.

Phase 3 Full-Scale Test Conclusions

- Thermophilic sludge dewatered better, producing higher cake solids (23% to 27%) and solids capture efficiency (84% to 94%), than mesophilic sludge (cake solids: 20% to 22%, solids capture efficiency: 66% to 88%) in full-scale tests at the EBMUD MWWTP. CST tests predicted the opposite, suggesting that the CST test may not be a good predictor of thermophilically digested sludge dewatering performance.
- VOCs were significantly higher in the full-scale thermophilic digester (4,000–6,000 ppbv at 50°C) compared to the full-scale mesophilic control digester (1,000–4,000 ppbv at 35°C). Sulfide levels in the thermophilic digester gas (512 ppm average) were somewhat higher than that the mesophilic digester gas (438 ppm average).
- Sentinel Chamber tests showed that all helminth ova were killed after being held for 4 hr in the full-scale digester operated at 50°C, but there was very little ova kill at 45°C after 12 hr.
- The full-scale thermophilic digester consistently met Class A fecal coliform requirements when operated at 50°C.
- The full-scale thermophilic digester consistently met U.S. EPA vector attraction reduction (VAR) requirement of 38% or more volatile solids reduction.
- The full-scale thermophilic digester occasionally exceeded the TVA/TALK ratio of the mesophilic control but did not typically exceed 0.10 (generally accepted as stable operation for mesophilic digesters).

Data Analysis Conclusions

- Regression data from single-stage digesters all operated at a 15-day MCRT (from Phase 1 data) showed that temperature and pH had the most influence on biosolids fecal coliform densities. This suggests that another disinfecting agent that becomes more lethal as pH rises may play a role in reducing fecal coliform concentrations. This disinfecting agent may be ammonia, since the un-ionized ammonia fraction of measured ammonia-nitrogen is toxic and increases with increasing pH.

- Regressing the single-stage Phase 1 data showed a strong relationship between increasing TVA with increasing thermophilic temperatures. This is consistent with specific VFA concentrations measured in Phase 1, where propionic and other acid levels increased as thermophilic temperatures rose to 53°–62°C.
- Regression analysis suggests a weak relationship between digester sludge TALK and temperature. TALK decreases with increasing thermophilic temperature in single-stage and methane-phase digesters.
- Regression analysis suggests that there is no relationship between digester sludge pH and thermophilic digester temperature.

CHAPTER 1.0

INTRODUCTION

1.1 Background

Land application of biosolids that meet only U.S. EPA Class B pathogen standards (40 CFR 503) can be controversial because of the potential for Class B biosolids to contain human pathogens. More and more counties are banning Class B biosolids land application. Wastewater treatment agency managers are concerned about the future of biosolids land application and are considering treatment upgrades to achieve U.S. EPA Class A standards, which require the virtual elimination of pathogens from biosolids. While meeting these standards would significantly broaden the range of opportunities for biosolids recycling, the processes required to achieve Class A pathogen levels are often cost prohibitive. Low-cost processes for meeting Class A standards would provide attractive biosolids management alternatives. This research specifically examined thermophilic anaerobic digestion processes as alternatives to meet Class A standards. Presently, thermophilic anaerobic digestion is not classified as a Process to Further Reduce Pathogens (PFRP) under the 40 CFR Part 503 regulations (Standards for the Use or Disposal of Sewage Sludge, U.S. EPA, 1994).

At the beginning of this study, an extensive literature review was conducted to assess state-of-the-art thermophilic anaerobic digestion methodology. This literature review is summarized in Section 1.2, which follows. A complete list of references compiled as part of the literature review, including works not mentioned in the summary, is included as Section 1.3. Another list of references for works cited elsewhere in this report is included at the end of the report.

1.2 Literature Review Summary

Anaerobic digestion has long been, and continues to be, one of the most widely used processes for the stabilization of primary and secondary municipal wastewater treatment sludges. The great majority of past and current applications of anaerobic digestion to wastewater sludges have been in the mesophilic temperature range from 35°–40° C (95°–104° F) (Zinder, 1986). Anaerobic sludge digestion in the thermophilic temperature range from 45-65° C (113°–149° F) has been practiced to only a limited extent (Ghosh, 1986; Mitsdorffer, 1991; Zinder, 1986).

Reasons for the limited use of anaerobic digestion at temperatures higher than mesophilic include higher energy requirements for heating, reports of poor process stability, increased odor, and lower quality supernatant (filtrate/centrate) (Buhr and Andrews, 1977; Zinder, 1986). The advantages of thermophilic anaerobic digestion over mesophilic anaerobic digestion have, in the past, accrued from increased stabilization and methane production rates and from improvements in sludge dewatering properties. Since the advent of the 40 CFR Part 503 Regulations, more studies have focused on the destruction of pathogenic organisms (Aitken and Mullenix, 1992; Han et al., 1997; Han and Dague, 1997; Huyard et al., 1998; Volpe et al., 1993; Watanbe et al.,

1997). Thermophilic anaerobic digestion has the significant added advantage of improved pathogen destruction with the potential for meeting the pathogen quality requirements of U.S. EPA Class A biosolids. While the disadvantages of thermophilic anaerobic digestion may have outweighed the advantages of the process in the past, U.S. EPA 503 regulations may well alter this balance.

The production of Class A stabilized solids weighs heavily when considering the total cost of sludge disposal. Since thermophilic anaerobic digestion can satisfy the pathogen reduction requirements for Class A biosolids, the economics of sludge processing may swing in favor of thermophilic digestion. Recent applications of thermophilic digestion in Germany and in the U.S. show that when properly designed and operated, thermophilic anaerobic digestion of municipal sludges can be carried out in a stable, nuisance-free, energy-efficient fashion.

The thermophilic anaerobic digestion process instability, which is often cited in the literature and given by U.S. EPA as the reason for excluding thermophilic anaerobic digestion as a PFRP, seems to be related to a particular method of process start-up. Most early workers used a "slow start-up method" to initiate thermophilic conditions from a mesophilically digesting sludge. This method seems to produce a microbial culture of thermotolerant mesophiles living at the edge of their survival range and very sensitive to temperature fluctuations (leading to process instability). The "fast start-up method" is now favored and produces a much more robust thermophilic culture (Aitken and Mullenix, 1992; Zinder, 1986).

Recent theoretical and experimental work has shown that draw/fill operation (digested sludge withdrawal prior to feed sludge addition) will produce significantly higher reductions of bacterial densities than fill/draw operation (feed sludge addition prior to digested sludge withdrawal) (Farrell et al., 1988). Even though draw/fill operation improves pathogen removal, most anaerobic digesters employ fill/draw or continuous feeding and withdrawal protocols to promote uniform gas production. It can also be shown theoretically that the efficiency of bacterial density reduction in a reactor is related to the frequency of feeding. Low frequency gives higher efficiency, but because low feeding frequency decreases the stability of digestion, a balance must be sought between these two objectives.

Several studies performed over the past two decades suggest that multiple-stage digestion has an advantage over a single-stage system with respect to pathogen reduction (Ghosh, 1986, 1987; Ghosh and Buoy, 1993; Ghosh et al., 1975; Ghosh et al., 1987; Ghosh et al., 1995; Lee et al., 1989; Mitsdorffer et al., 1990; Wu, 1987). Furthermore, these studies suggest that some of the reported disadvantages of thermophilic anaerobic digestion of municipal sludges can be eliminated if it is used in a two-stage, two-phase digestion system where the first-stage is thermophilic and the second-stage is mesophilic. Such a system is advantageous for pathogen removal in several ways:

- It provides for less short-circuiting than a single stage system, especially if the first stage can be operated in an infrequently-fed draw/fill mode, rather than the typical fill/draw sequence;

- The second stage can produce a typical mesophilically-digested sludge in terms of odor and supernatant (filtrate/centrate) quality. Such a system requires a tightly sealed first-stage and interstage transfer piping to contain odors; and
- A phased system may more efficiently digest organic material (e.g., higher rate of hydrolysis and acetogenesis).

To summarize, studies reviewed in the literature review described some improvements to the stability and performance of the thermophilic anaerobic digestion process. These improvements were incorporated into the processes evaluated as part of this research and include:

- Fast start-up method. The fast start-up method produces a more robust thermophilic culture that is tolerant of temperature fluctuations and increases process stability.
- Draw/Fill operation. The Draw/Fill cycle improves pathogen removal by reducing short-circuiting.
- Multiple-stage digestion. Multiple-stage reactors reduce short-circuiting and, if the second-stage reactor is mesophilic, produce a sludge of higher quality in terms of odor and supernatant. Phased systems may more efficiently digest organic material.

Pathogen destruction in municipal sludge digestion follows a typical time/temperature relationship, with higher temperatures requiring shorter exposure times for pathogen destruction. Data on pathogen survival in municipal sludge digestion suggest that thermophilic digestion achieves approximately two orders of magnitude higher pathogenic bacteria reduction than mesophilic digestion and is a good candidate for meeting the pathogenic bacteria levels required for Class A sludge (Aitken and Mullenix, 1992; Berg and Berman, 1980; Garber, 1977, 1982; Lee et al., 1989; Popova and Bolotina, 1964; Stern and Farrell, 1977; Volpe et al., 1993). U.S. EPA studies (Lee et al., 1989). Full-scale research results (Ghosh and Buoy, 1993) indicate enhanced pathogen destruction in acid-phase digestion.

Tables 1-1 and 1-2 summarize the thermophilic anaerobic digestion experiments and applications reviewed in preparation for this project. In addition, we referred to review papers (Buhr and Andrews, 1977; Kotze et al., 1969; Stern and Farrell, 1977; Zinder, 1986), regulatory documents published by the U.S. EPA (U.S. EPA, 1989, 1994) and the German Abwasser Technische Vereinigung (ATV) Working Group 3.2.2 on Disinfection of Sewage Sludge, and relevant chapters of engineering texts (Ahring and Westermann, 1988; Farrell and Southworth, 1992; Metcalf & Eddy, 1991; WEF and ASCE, 1991). A complete listing of technical papers collected as part of the literature review is provided as Section 1.3.

Table 1-1. Plant-Scale Experiments/Applications of Thermophilic Anaerobic Digestion of Municipal Wastewater Sludge.

Location	Year	Anaerobic Digestion Process Configuration	Size of Experiment/ Application	Reference
Aurora, IL, USA	1932	Single-stage thermophilic	1 digester (1,200 m^3)	17
Jackson, MI, USA	1942 -1944	Three-stage thermophilic-thermophilic-mesophilic, and three-stage thermophilic-thermophilic-thermophilic	3 digesters (total capacity of 3 stages = 4,600 m^3)	17
Hyperion Treatment Plant Los Angeles, CA, USA	1952 - ?	Single-stage thermophilic	1 to 6 digesters (9,600 m^3/digester)	18-23
Kur'yanovo Treatment Plant Moscow, Russia	1955 - ?	Two-stage thermophilic-thermophilic	8 digesters (5,200 m^3/digester)	54
West-Southwest Sewage Treatment Works Chicago, IL, USA	1978 -1980	Single-stage thermophilic	1 digester (8,900 m^3)	55
Osterode, Germany	1987	Two-stage thermophilic-mesophilic	2 - 1st-stage digesters (130 m^3/digester) and 1 - 2nd-stage digester (900 m^3)	50
Woodridge-Greenevalley Municipal Water Reclamation Plant, Du Page County, IL USA	1990 - Present	Two-stage mesophilic-thermophilic	1 acid-phase digester (300 m^3) and 1 methane-phase digester (1850 m^3)	24, 25, and 26
Greater Vancouver Regional District Lion's Gate Plant, Vancouver, BC, Canada	1990 - Present	Two-stage thermophilic-thermophilic	1 - 1st stage digester (155 m^3) and 1 - 2nd stage digester (155 m^3)	69
Geseke, Germany	1991	Two-stage thermophilic-mesophilic	1 - 1st stage digester (130 m^3) and 1 - 2nd stage digester (1,600 m^3)	49
Altenmarkt, Germany	1991	Two-stage thermophilic-mesophilic	1 - 1st stage digester (60 m^3) and 1 - 2nd stage digester (300 m^3)	49
Cologne, Germany	1993	Two-stage thermophilic-mesophilic	NA	13
Greater Vancouver Regional District Anacis Plant, Vancouver, BC, Canada	1996- Present	Two-stage thermophilic-thermophilic	1- 1st stage digester (15 d HRT 1- 2nd stage digester (1-1.5 d HRT)	69
Papillion Creek Plant, Omaha, NE, USA	1995?	Two-stage thermophilic-mesophilic	1-1st stage digester (7-10 d HRT) 2- 2nd stage digester (13-20 d HRT)	65

Table 1-1. Plant-Scale Experiments/Applications of Thermophilic Anaerobic Digestion of Municipal Wastewater Sludge, continued.

Location	Year	Anaerobic Digestion Process Configuration	Size of Experiment/ Application	Reference
Newton, IA, USA	1995	Two-stage thermophilic-mesophilic	1- 1st stage digester (30 d HRT) 2- 2nd stage digester (60 d HRT)	43
Sturgeon Bay Utilities, WI, USA	1996	Two-stage thermophilic-mesophilic	1- 1st stage digester (25-30 d HRT) 1- 2nd stage digester (30 d HRT)	68
Independence, IA, USA	1988	Two-stage thermophilic-mesophilic	1-1st stage digester (3800 m^3) (~65 d HRT) 2- 2nd stage digester (1140 m^3 each) (40 d HRT)	11
Neenah-Menasha Sewerage Commission, Milwaukee, WI, USA	1998	Two-stage thermophilic-mesophilic	1- 1st stage digester (2700 m^3 gas) (16 d HRT) 2- 2nd stage digester (2700 m^3 each gas) (16 d HRT)	51

Table 1-2. Laboratory-Scale Experiments on Thermophilic Anaerobic Digestion of Municipal Wastewater Sludge.

Researcher (Year)	Focus of Experiment	Anaerobic Digestion Process Configuration	Sludge Used in Experiments	Reference
Heukelekian and Kaplovsky (1948)	Determine effects of abrupt temperature changes during anaerobic digestion on the microbial thermophilic population	Single-stage mesophilic and thermophilic	New Brunswick, NJ, USA	36
Golueke (1958)	Investigate effect of temperature on a variety of parameters used to evaluate the anaerobic digestion performance	Single-stage mesophilic and thermophilic	Not mentioned (experiment performed at University of California at Berkeley)	31
Malina (1961)	Evaluate the effect of temperature on high-rate digestion of waste activated sludge	Single-stage mesophilic and thermophilic	Nine Springs Sewage Treatment Plant Madison, WI, USA	44
Maly and Fadrus (1971)	Evaluate influence of temperature on fermentation rate and gas production of anaerobic digestion	Single-stage mesophilic and thermophilic	Wastewater Treatment Plant, Brno, Czechoslovakia	46

Table 1-2. Laboratory-Scale Experiments on Thermophilic Anaerobic Digestion of Municipal Wastewater Sludge, continued.

Researcher (Year)	Focus of Experiment	Anaerobic Digestion Process Configuration	Sludge Used in Experiments	Reference
Berg and Berman (1980)	Investigate destruction of indicator bacteria and viruses during thermophilic anaerobic digestion	Single-stage mesophilic and thermophilic	Hyperion Treatment Plant Los Angeles, CA, USA	8
Wu (1987)	Compare performance of single-stage anaerobic digester with three-stage anaerobic digestion and thermal treatment	Hydrolyzing mesophilic stage, followed by an acid-forming mesophilic stage, followed by a methane-forming thermophilic stage	Allegheny County Sanitary Authority, Pittsburgh, PA, USA	74
Huyard et al. (1998)	Determine inactivation of pathogens by thermophilic acid plus mesophilic anaerobic digestion	Two-stage thermophilic acid and mesophilic methane digestion.	Primary sludge and waste activated sludge from Belmont, Indianapolis, IN.	37
Ekola (1998)	Determine fecal coliform and salmonella reduction by two-stage thermophilic plus mesophilic digester. First thermophilic stage is most likely acid.	Two stage thermophilic (acid?) anaerobic digestion plus mesophilic anaerobic digestion.	Waste activated sludge (no primary clarifiers)	14
Farrell et al. (1988)	Determine effects of feeding protocol on microbial densities and other process parameters of a single-stage digester	Single-stage mesophilic	Mill Creek Wastewater Treatment Plant Cincinnati, OH, USA	16
Lee et al. (1989)	Compare pathogenic microorganism destruction obtained during single-stage and two-phase (mesophilic and thermophilic) anaerobic digestion	Two-stage. Both stages operated at mesophilic and thermophilic temperatures	Cincinnati Metropolitan Sewer District Mill Creek Plant Cincinnati, OH, USA	42
Medhat and Stafford (1989)	Determine fate of *Entamoeba histolytica* during mesophilic and thermophilic digestion	Not mentioned	Cardiff, UK	47
Aoki and Kawase (1991)	Compare single-stage mesophilic and thermophilic anaerobic digestion with high-performance thermophilic two-phase digestion (HPTTD)	HPTTD consists of: 1) thermal sludge conditioning at 90°C, 2) high temperature (70°C) acid fermentation, and 3) fixed-bed reactor methane fermentation at 55°C	Wastewater Treatment Plant in Maegata-cho, Handa, Japan	5

Table 1-2. Laboratory-Scale Experiments on Thermophilic Anaerobic Digestion of Municipal Wastewater Sludge, continued.

Researcher (Year)	Focus of Experiment	Anaerobic Digestion Process Configuration	Sludge Used in Experiments	Reference
Aitken and Mullennix (1992)	Evaluate indicator organism destruction capability of thermophilic anaerobic digestion and stability of the thermophilic population	Single-stage thermophilic	Orange Water and Sewer Authority Mason Farm Treatment Plant Chapel Hill, NC, USA	4
Ghosh et al. (1993 and 1995)	Demonstrate efficiency of two-phase anaerobic digestion for improved solids stabilization and elimination of digester foaming	Two-stage mesophilic-mesophilic, and two-stage mesophilic-thermophilic	Woodridge-Greenevalley Municipal Water Reclamation Plant DuPage County, IL, USA	24, 25, and 26
Volpe et al. (1993)	Measure pathogen die-off rates during thermophilic anaerobic digestion	Two-stage thermophilic-thermophilic	Greater Vancouver Regional District (GVRD) Lion's Gate Plant, Vancouver, BC, Canada	69
Han and Dague (1997)	Demonstrate ability of temperature-phased anaerobic digestion (TPAD) to reduce fecal coliforms. Examine process efficiency.	Two-stage thermophilic-mesophilic	Water Pollution Control Plant, Ames, IA (primary sludge)	32
Han et al. (1997)	Demonstrate ability of TPAD to reduce fecal coliforms. Examine process efficiency to reduce indicated organisms.	Two-stage thermophilic-mesophilic	Water Pollution Control Plant, Ames, IA (primary sludge plus waste activate sludge)	33
Watanabe et al. (1997)	Studiy the performance of thermo and meso anaerobic digestion to reduce pathogens.	Single stage mesophilic and thermophilic	Primary sludge from a full-scale plant in Japan, exact location unknown	72

1.3 References for Literature Review

1. Ahring, B.K. and Westermann, P. (1988) Toxicity of Heavy Metals to Thermophilic Anaerobic Digestion. In: *Biotreatment Systems (Volume II).* D.L. Wise (Ed.), CRC Press, Boca Raton, FL, 1.
2. Ahring, B.K. and Westermann, P. (1985) Sensitivity of Thermophilic Methanogenic Bacteria to Heavy Metals. *Curr. Microbiol.*, **12**, 273.
3. Ahring, B.K. and Westermann, P. (1983) Toxicity of Heavy Metals to Thermophilic Anaerobic Digestion. *Eur. J. Appl. Microbiol. Biotechnol.*, **17**, 365.
4. Aitken, M.D. and Mullennix, R.W. (1992) Another Look at Thermophilic Anaerobic Digestion of Wastewater Sludge. *Water Environ. Res.*, **64**, 915.

5. Aoki, N. and Kawase, M. (1991) Development of High-Performance Thermophilic Two-Phase Digestion Process. *Water Sci. Technol.*, **23**, 1147.
6. ATV/VKS Working Group 3.2.2. on Disinfection of Sewage Sludge. First Report (1986). ATV/VKS Working Group 3.2.2. on Disinfection of Sewage Sludge. Second Report (1988).
7. Beall, S.S., Jenkins, D., and Vidanage, S. (1998) A Systematic Analytical Artifact That Significantly Influences Anaerobic Sludge Digestion Efficiency Measurement. *Water Environ. Res.* **70**:1019-1024.
8. Berg, G. and Berman, D. (1980) Destruction by Anaerobic Mesophilic and Thermophilic Digestion of Viruses and Indicator Bacteria Indigenous to Domestic Sludges. *Appl. Environ. Microbiol.*, **39**, 361.
9. Bertucci, J.J., Lue-Hing, C., Zenz, D., and Sedita, S.J. (1977) Inactivation of Viruses during Anaerobic Sludge Digestion. *J. Water Pollut. Control Fed.*, **49**, 1642.
10. Buhr, H.O. and Andrews, J.F. (1977) The Thermophilic Anaerobic Digestion Process. *Water Res.*, **11**, 129.
11. Chudzinski, K. (1988) Privately Communicated, Independence, IA.
12. Deitsch, J. (1998) Privately Communicated, Milwaukee, WI.
13. Dichtl, N. (1997) Thermophilic and Mesophilic (Two-Stage) Anaerobic Digestion. *J. Chartered Inst., Water Environ. Management,* **11**, 98.
14. Ekola, T. (1998) Privately Communicated, Western Lake Superior Sanitary District, Duluth, MN.
15. Farrell, J.B. and Southworth, R.M. (1992) Regulatory Issues. Chapter 2 In: *Municipal Sewage Sludge Management (Water Quality Management Library - Volume 4).* C. Lue-Hing, D.R. Zenz, and R. Kuchenrither (Eds.), Technomic Publishing Co., 3.
16. Farrell, J.B., Erlap, A.E., Rickabaugh, J., Freedman, D., and Hayes, S. (1988) Influence of Feeding Procedure on Microbial Reductions and Performance of Anaerobic Digestion. *J. Water Pollut. Control Fed.*, **60**, 635.
17. Fischer, A.J. and Greene, R.A. (1945) Plant Scale Tests on Thermophilic Digestion. *Sewage Works J.*, **17**, 718.
18. Garber, W.F. (1982) Operating Experience with Thermophilic Anaerobic Digestion. *J. Water Pollut. Control Fed.*, **54**, 1170.
19. Garber, W.F. (1977) Certain Aspects of Anaerobic Digestion of Wastewater Solids in the Thermophilic Range at the Hyperion Treatment Plant. *Prog. Wat. Technol.*, **8**, 401.
20. Garber, W.F., Ohara, G.T., Raksit, S.K., and Olson, D.R. (1977) Studies of Dewatering Anaerobically Digested Wastewater Solids at the Hyperion Treatment Plant. *Prog. Wat. Technol.*, **8**, 371.
21. Garber, W.F., Ohara, G.T., Colbaugh, J.E., and Raksit, S.K. (1975) Thermophilic Digestion at the Hyperion Treatment Plant. *J. Water Pollut. Control Fed.*, **47**, 950.
22. Garber, W.F., Ohara, G.T., and Raksit, S.K. (1975) Energy-Wastewater Treatment and Solids Disposal. *J. Environ. Eng. Div.*, ASCE, **101**, 319.
23. Garber, W.F. (1954) Plant-Scale Studies of Thermophilic Digestion at Los Angeles. *Sewage Ind. Wastes*, **26**, 1202.

24. Ghosh, S., Buoy, K., Dressel, L., Miller, T., Wilcox, G., and Loos, D. (1995) Pilot- and Full-Scale Two-Phase Anaerobic Digestion of Municipal Sludge. *Water Environ. Res.*, **67**, 206.
25. Ghosh, S. and Buoy, K. (1993a) The Acimet Process: An Innovative Approach to Biogasification of Municipal Sludge. In: *Proc. 1st Biomass Conference of the Americas.* Burlington, VT, 759.
26. Ghosh, S. and Buoy, K. (1993b) Improved Sludge Stabilization and Digester Foam Control by the Acimet Process. *Abstract of Paper for Presentation at the Utah WPCA Annual Meeting*, Park City, UT.
27. Ghosh, S. (1987) Improved Sludge Gasification by Two-Phase Anaerobic Digestion. *J. Environ. Eng.*, **113**, 1265.
28. Ghosh, S., Henry, M.P., and Sajjad, A. (1987*) Stabilization of Sewage Sludge by Two-Phase Anaerobic Digestion*. U.S. EPA/600/S2-87/040. U.S. EPA, Cincinnati, OH.
29. Ghosh, S. (1986) Comparative Studies of Temperature Effects on Single-Stage and Two-Phase Anaerobic Digestion. *Eighth Symposium on Biotechnology for Fuels and Chemicals (Biotech, Bioengrg. Symp. No. 17)*, Interscience Publication by John Wiley & Sons, New York, NY, 365.
30. Ghosh, S., Conrad, J.R., and Klass, D.L. (1975) Anaerobic Acidogenesis of Wastewater Sludge. *J. Water Pollut. Control Fed.*, **47**, 30.
31. Golueke, C.G. (1958) Temperature Effects on Anaerobic Digestion of Raw Sewage Sludge. *Sewage Ind. Wastes*, **30**, 1225.
32. Han, Y. and Dague, R.R. (1997) Laboratory Studies on the Temperature-phased Anaerobic Digestion of Domestic Primary Sludge. *Water Environ. Res.*, **69**, 1139.
33. Han, Y., Sung, S. and Dague, R.R. (1997) Temperature-Phased Anaerobic Digestion of Wastewater Sludges. *Water Sci. Technol.*, **36**, 367.
34. Hammer, M.S. and Borchardt, J.A. (1969) Dialysis Separation of Sewage Sludge Digestion. *J. San. Eng. Div., Proc. ASCE*, **95**, 907.
35. Haug, R.T. (1977) Sludge Processing to Optimize Digestibility and Energy Production. *J. Water Pollut. Control Fed.*, **49**, 1713.
36. Heukelekian, H. and Kaplovsky, A.J. (1948) Effect of Change of Temperature on Thermophilic Digestion. *Sewage Works J.*, **20**, 806.
37. Huyard, A., Ferran, B., Audic, J.M., Dieterien, J., Adamik, J., and Noel, T. (1998) A Challenge for the Two-Phased Anaerobic Digestion: To Produce Class A Biosolids and Meet PFRP Equivalency. Presented at WEFTEC 98, Orlando, FL. Oct. 1998.
38. Kabler, P. (1959) Removal of Pathogenic Microorganisms by Sewage Treatment Processes. *Sewage Ind. Wastes*, **31**, 1373.
39. Kabrick, R.M. and Jewel, W.J. (1982) Fate of Pathogens in Thermophilic Aerobic Sludge Digestion. *Water Res.*, **16**, 1051.
40. Kelly, S. and Sanderson, W.W. (1959) The Effect of Sewage Treatment on Viruses. *Sewage Ind. Wastes*, **31**, 683.
41. Kotze, J.P., Thiel, P.G., and Hattingh, W.H.J. (1969) Anaerobic Digestion II – The Characterization and Control of Anaerobic Digestion. *Water Res.*, **3**, 459.

42. Lee, K.M., Brunner, C.A., Farrell, J.B., and Eralp, A.E. (1989) Destruction of Enteric Bacteria and Viruses during Two-Phase Digestion. *J. Water Pollut. Control Fed.*, **61**, 1421.
43. Main, R.E. (1998) Privately Communicated, Newton, IA.
44. Malina, J.F. Jr. (1961) The Effect of Temperature on High-Rate Digestion of Activated Sludge. In: *Proc. 16th Ind. Waste Conf.,* Purdue University, Lafayette, IN, 232-250.
45. Marina, T. (1998) Privately Communicated, McMahon and Assoc. Milwaukee, WI.
46. Maly, J. and Fadrus, H. (1971) Influence of Temperature on Anaerobic Digestion. *J. Water Pollut. Control Fed.*, **43**, 641.
47. Medhat, A.E.F. and Stafford, D.A. (1989) The Fate of *Entamoeba histolytica* During Sewage Treatment Using Anaerobic Digesters. In: *Watershed 89: The Future for Water Quality in Europe (Vol II),* 381.
48. Metcalf & Eddy, Inc. (1991) *Wastewater Engineering: Treatment, Disposal, and Reuse (3rd ed.).* McGraw-Hill, New York, NY, 425.
49. Mitsdorffer, R. (1991) Charakteristika der Zweistufagen Thermophilen/Mesophilen Schlammfaulung unter Berucksichtigung Kinetischer Ansatze. Ber. Wassergute und Abfallwirtschaft, Technische Univ., Munchen, Germany, No. 109.
50. Mitsdorffer, R., Demharter, W., and Bischofsberger, W. (1990) Stabilization and Disinfection of Sewage Sludge by Two-Stage Anaerobic Thermophilic/Mesophilic Digestion. *Wat. Sci. Technol.*, **22**, 289.
51. Much, R. (1998) Private Communicated. Neenah-Menasha Sewerage Commission, Milwaukee, WI.
52. Oles, J. (1995) Privately Communicated, and Oswald Schulze GmbH and Co. KG Brochure "Sludge Digestion Thermophilic/Mesophilic."
53. Pohland, F.G. and Ghosh, S. (1971) Developments in Anaerobic Treatment Processes. In: *Biological Waste Treatment*. R. P. Canale (Ed), Wiley Interscience, New York.
54. Popova, N.M. and Bolotina, O.T. (1964) The Present State of Purification of Town Sewage and the Trend in Research Work in the City of Moscow. In*: Advances in Water Pollution Research (Vol. 2).* W.W. Eckenfelder (Ed.), Macmillan, New York, 97.
55. Rimkus, R.R., Ryan, J.M., and Cook, E.J. (1982) Full-Scale Thermophilic Digestion at the West-Southwest Sewage Treatment Works, Chicago, Illinois. *J. Water Pollut. Control Fed.*, **54**, 1447.
56. Rudolfs, W. and Heukelekian, H. (1930) Thermophilic Digestion of Sewage Solids—I. Preliminary paper. *Ind. Engng Chem.*, **22**, 96.
57. Schaumburg, F.D. and Kirsch, E.J. (1966) Anaerobic Simulated Mixed Culture System. *Appl. Microbiol.*, **14**. 761.
58. Shih, J.C.H. (1988) Pathogen Control by Anaerobic Digestion. In: *Anaerobic Digestion 1988*. Pergamon Press, Oxford, U.K., 259.
59. Smith, P.H. and Mah, R.A. (1966) Kinetics of Acetate Metabolism During Sludge Digestion. *Appl. Microbiol.*, **14**, 368.
60. Steiner, A. (1986) Schlammstabilisierung unter dem Aspekt der

Klarschlammverordnung - Hygiene Inst. fur Wasserwesen, Hochschule der Bundeswehr, Munchen, Germany, Mitteilunger, 16.

61. Stern, G. and Farrell J.B. (1977) Sludge Disinfection Techniques. In: *Proc. Nat. Conf. on Composting of Municipal Residues and Sludges*. Pub. Information Transfer, Inc., Rockville, MD,142.
62. Strauch, D. and DeBertoldi, M. (1986) Microbiological Specifications of Sewage Sludge in Processing and Use of Organic Sludge and Liquid Organic Wastes, Rome, Italy.
63. Streeter, M.D., Dague, R.R., and Main, R.E. Evaluation of a Field Application, Temperature Phased Anaerobic Digestion. Presented at WEFTEC'97, Chicago, IL. Oct. 1997.
64. Streeter, M.D. (1996) Evaluation of a Field Application of the Temperature Phased Anaerobic Digestion Process. M.S. Thesis, Civil (Environmental) Engineering, Iowa State University, Ames, IA.
65. Swan, T. (1998) Privately Communicated, Papillion Creek Wastewater Treatment Plant, Omaha, NE.
66. U.S. EPA (1994) *Standards for the Use or Disposal of Sewage Sludge*. 40 CFR Part 503 (revised July 1, 1994).
67. U.S. EPA (1989) *Control of Pathogens in Municipal Wastewater Sludge*. U.S. EPA Pub. No. 625/10-89/006. U.S. EPA Center for Environmental Research Information, Cincinnati, OH.
68. Vik, T.E. and Olsen, C.T. (1997) Full-Scale Operation of Temperature-Phased Anaerobic Digestion. Presented at Wisconsin Wastewater Operator's Assoc., La Crosse, WI, Oct. 1997.
69. Volpe, G., Rabinowitz, B., Peddie, C., and Krugel, S. (1993) Class A (High Grade) Sludge Process Design for the Greater Vancouver Regional District (GVRD) Annacis Island Wastewater Treatment Plant. In: *Proc. Water Env. Fed. 66th Annual Conf. and Expo. (Sludge Management Vol. 4.)*, Anaheim, CA, 1.
70. Ward, R.L. and Ashley, C.S. (1978) Methods to Inactivate Enteric Viruses in Wastewater Sludge. Sandia Laboratories (Sand77-1498). Albuquerque, NM, 1.
71. Ward, R.L., Ashley, C.S., and Moseley, R.H. (1976) Heat Inactivation of Poliovirus in Wastewater Sludge. *Appl. Environ. Microbiol.*. **32**, 339.
72. Watanabe, H., Kitamura, T., Ochi, S. and Ozaki, M. (1997) Inactivation of Pathogenic Bateria Under Mesophilic and Thermophilic Conditions. *Water Sci. Technol.*, **36**, 25.
73. Water Environment Federation and American Society of Civil Engineers. (1991) *Design of Wastewater Treatment Plants (Volume II).* WEF Manual of Practice No. 8 and ASCE Manual and Report on Engineering Practice No. 76, 1272.
74. Wu, Y.C. (1987) *Destruction and Stabilization of Sludge by Multiple-Stage Digestion and Thermal Treatment*. EPA/600/S2-86/118. U.S. EPA, Cincinnati.
75. Zinder, S.H. (1986) Thermophilic Waste Treatment Systems. In: *Thermophiles*. T.D. Brock (Ed.), John Wiley & Sons, New York, 257.

1.4 Project Scope

The objectives of this study were to:

- Evaluate and compare candidate thermophilic anaerobic digestion processes for meeting Class A requirements;
- Apply for a Class A pathogen equivalency (PFRP) recommendation from the U.S. EPA Pathogen Equivalency Committee (PEC); and
- Evaluate other aspects of thermophilic anaerobic digester performance including process stability, dewaterability, and odor generation.

The effectiveness of thermophilic processes to achieve Class A was evaluated using U.S. EPA Alternatives 3 and 4 (40 CFR 503 Section 503.32(a)(5) and (6), demonstration of enteric virus and helminth ova reduction), rather than Alternative 1 (Section 503.32(a)(3), meeting time/temperature requirements).

The research was divided into three phases. In Phase 1, three thermophilic anaerobic digestion process configurations were screened at bench-scale and evaluated using fecal coliform data. These processes were:

- Thermophilic acid phase;
- One-stage thermophilic; and
- Two-phase, mesophilic acid phase/thermophilic methane phase.

Each process configuration was examined at three thermophilic temperatures: approximately 48° C, 53° C, and 62° C, and compared with a mesophilic control at 35° C. Processes showing potential for meeting the Class A fecal coliform requirement (<1000 MPN/g TS) at one or more temperatures, would be carried forward for further evaluation in Phase 2, where the feed sludge was seeded with pathogens and digester performance was evaluated on the basis of both fecal coliform and pathogen destruction. Those processes that met Class A requirements in Phase 2 were considered for evaluation in the Phase 3, full-scale pilot study. The objectives of the full-scale test were to confirm bench-scale pathogen destruction results and evaluate other aspects of digester performance including solids stabilization, process stability, digested sludge dewaterability, and odor generation.

A goal of this study was to maintain a uniformity of conditions, particularly regarding feed sludge, which is not evident in much previously reported work. Although several studies have evaluated various thermophilic anaerobic digestion processes, few have been conducted at a single facility with a single feed sludge. Previous review papers have had difficulty in comparing process performance for systems operated at different facilities. Reviewers have noted that "…raw sludge characteristics vary widely. To some extent differences in the facilities' mixing and heating systems also affect performance" (Schafer and Farrell, 2000). Bench-scale studies conducted as part of this research directly addressed this issue by conducting parallel operation of different processes using identical equipment and feed sludges.

CHAPTER 2.0

PHASE 1 SCREENING EVALUATION

2.1 Purpose

The purpose of Phase 1 was to evaluate several configurations of the thermophilic anaerobic digestion process and to select one or more of these processes for Phase 2 pathogen testing. Selection was based primarily upon fecal coliform destruction efficiency. Process stability was also considered. Three thermophilic anaerobic digestion process configurations were screened at bench-scale: 1) two-stage thermophilic acid phase/mesophilic methane phase; 2) one-stage thermophilic; and 3) two-stage mesophilic acid phase/thermophilic methane phase. Each process was examined at three different thermophilic temperatures (about 48° C, 52° C, and 62° C) and compared against a mesophilic control at 35° C (referred to as Meso 35). Because the thermophilic acid stage in Process 1 was expected to provide significantly greater disinfection power than the mesophilic methane stage, only the acid stage was operated for evaluation purposes in Phase 1.

2.2 Materials and Methods

Laboratory-scale Plexiglas digesters with an operating capacity of approximately 30-L were used in Phase 1. The contents were mixed by an externally mounted variable-speed DC motor that turned a stainless steel shaft and two 4-in. diameter propellers. The mixing system was sealed, and anaerobic conditions were maintained by a small tube that covers the stainless steel shaft (Figure 2-1). Each digester was heated by a 520-watt heat tape (Cole Parmer, E-36110-08, Chicago, IL). To avoid direct contact between the Plexiglas reactor walls and heat tape, the exterior vertical area of the cylinders was covered with aluminum foil, which also distributed heat more evenly. Digester temperature was controlled by a proportional-response temperature controller (Dynasense Model 02149-20, W. Palm Beach, FL) that received a signal from an 18-in. submersible resistance temperature detector (RTD) probe (Cole-Parmer, E-08117-74, Chicago, IL). The temperature controller did not include a temperature display, and temperatures were measured using external bimetal thermometers during tests for Processes 1 and 2 to adjust heat tape output accordingly. During the Process 3 test, the bimetal thermometers were replaced with calibrated mercury thermometers for greater accuracy and precision. Gas production was measured using a wet-tip gas meter (Wet Tip Gas Meter Co., Nashville, TN).

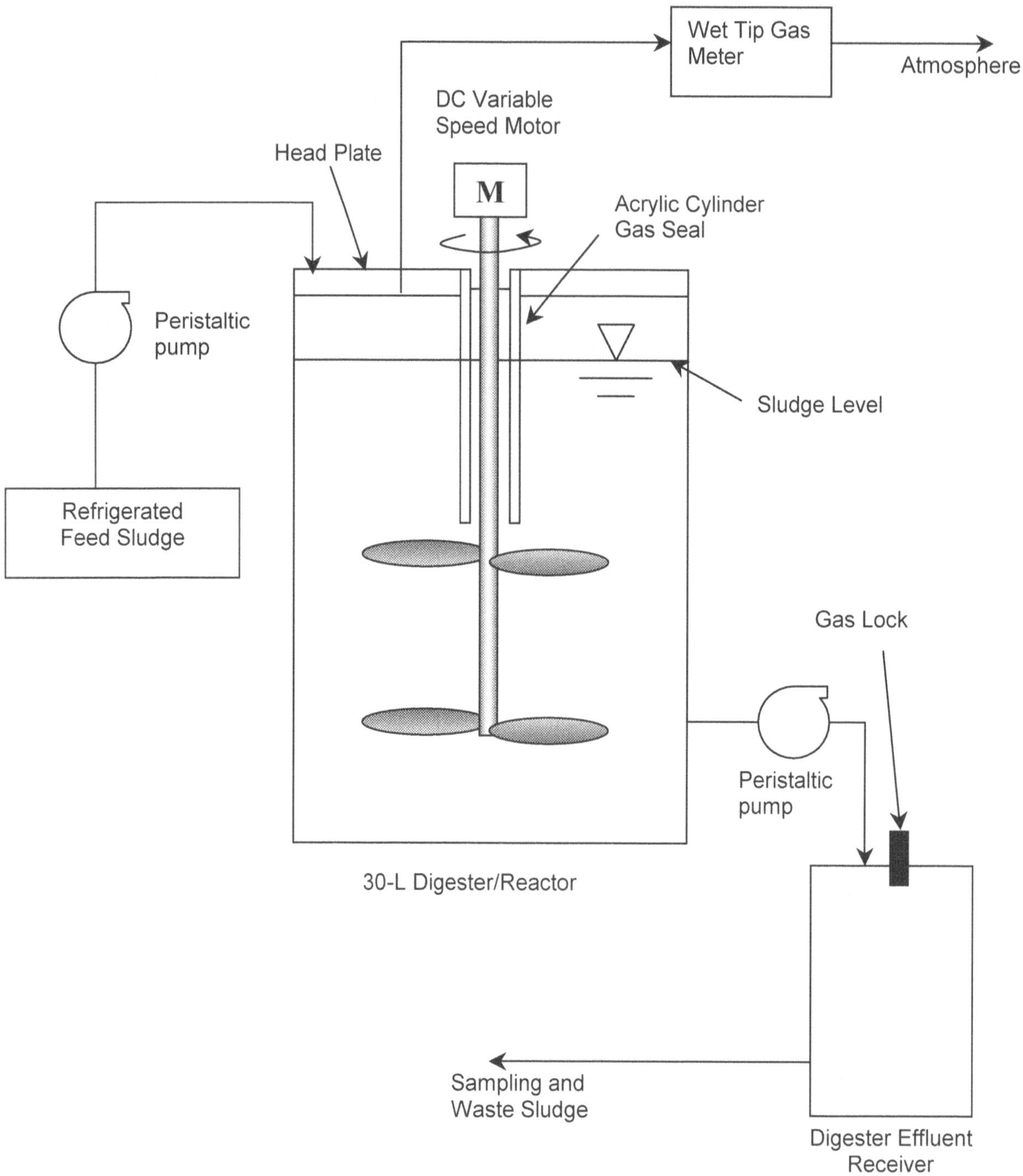

Figure 2-1. Process 1 Digester Schematic.

Peristaltic pumps (Masterflex, Cole-Parmer, Chicago, IL) controlled digester influent and effluent flows. The various residence times (2 days for acid-phase and 15 days for methane phase) were regulated by adjusting flow rates and pumping intervals. A semi-continuous flow regime (hourly feed cycles) was used for the acid-phase digesters to minimize temperature fluctuations due to their high feed rates. To further minimize temperature fluctuations in the acid-phase digesters, feed sludge was preheated (35°C). Due to the lower feed rates, feed cycles for single-stage and methane-phase digesters were 2–6 hours apart. Digester effluent was stored in a container with a gas seal that allowed gas to escape but prevented air from entering. A photograph of the Phase 1 laboratory-scale digester system is included in Appendix A (Figure A-1).

Digester feed was a mixture of EBMUD main wastewater treatment plant's primary sludge (PS) and thickened waste activated sludge (TWAS). The PS was filtered through a stainless steel wire mesh (5×5 mesh per lineal inch with 63.2% open area) to remove coarse solid material that could clog pumps and tubing and mixed in a 1:1 v/v ratio with TWAS. The feed was kept in a refrigerator at approximately 4°C and was mixed using a submerged pump and recirculation loop prior to feeding.

All laboratory tests were performed in accordance with procedures detailed in *Standard Methods for the Examination of Water and Wastewater 20th ed.* (American Public Health Association, 1998).

2.3 Results and Discussion

2.3.1 Process 1: Thermophilic Acid

The complete digestion process would be a thermophilic acid digester usually followed by a mesophilic methane digester. Because it was presumed that most of the disinfection would occur in the thermophilic acid phase digester, only this phase was tested in the screening evaluation (Figure 2-2). Subsequent tests of this process (see Phase 2 discussion in Chapter 3.0) included the mesophilic methane phase. Table 2-1 lists average values, standard deviations, and coefficients of variation for parameters measured for Process 1.

PROCESS 1

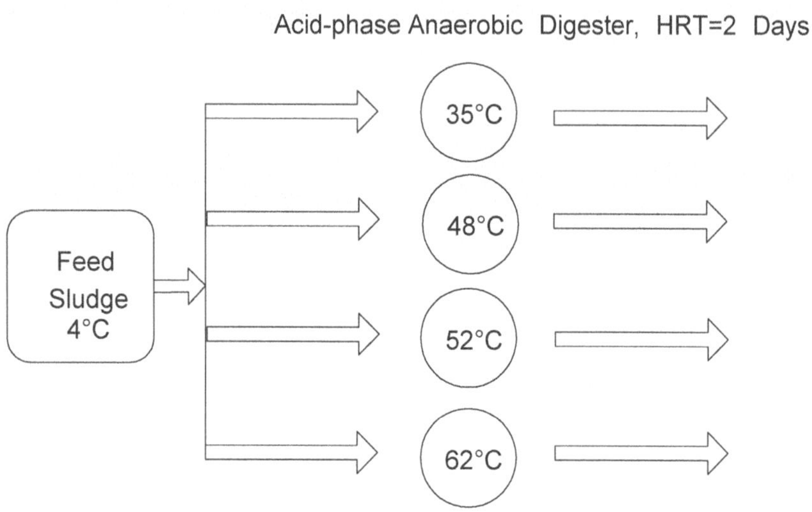

Figure 2-2. Phase 1 Process 1 Schematic.

Table 2-1. Process 1 Averages and Standard Deviations for Measured Parameters.

Parameter	Units	# of samples		Feed	Meso 35	Thermo 48	Thermo 54	Thermo 62
Residence time	days	53	average		3.4	2.5	2.8	2.7
			st. dev.		4.7	0.9	2.2	2.2
			coeff. of var.		1.4	0.4	0.8	0.8
Temperature	°C	105	average		35	48	54	62
			st. dev.		2	2	3	3
			coeff. of var.		0.05	0.04	0.06	0.05
Gas production	L/hr	54	average		0.50	0.03	0.80	0.21
			st. dev.		0.31	0.14	0.61	0.51
			coeff. of var.		0.61	4.91	0.77	2.44
Methane in gas produced	percent	5	average		24	24	33	34
			st. dev.		3	4	5	7
			coeff. of var.		0.12	0.16	0.16	0.21
pH		54	average	5.9	5.8	6.1	6.6	6.6
			st. dev.	0.1	0.2	0.1	0.1	0.2
			coeff. of var.	0.02	0.03	0.02	0.02	0.02
Total Alkalinity	mg $CaCO_3$/L	24	average		3400	4200	4900	5100
			st. dev.		500	500	600	500
			coeff. of var.		0.15	0.12	0.12	0.10
Total Volatile Acids	mg acetic acid/L	24	average		6700	7600	7300	7600
			st. dev.		700	800	700	800
			coeff. of var.		0.1	0.1	0.1	0.1
TVA/TALK ratio	mg acetic acid/mg $CaCO_3$	24	average		2.0	1.8	1.5	1.5
			st. dev.		0.2	0.2	0.2	0.1
			coeff. of var.		0.1	0.1	0.1	0.1
Acetic Acid	mg/L	7	average	1700	2800	3500	3200	3600
			st. dev.	200	400	300	400	600
			coeff. of var.	0.1	0.1	0.1	0.1	0.2
Propionic Acid	mg/L	7	average	900	1800	1400	1500	1700
			st. dev.	200	300	200	300	200
			coeff. of var.	0.2	0.2	0.1	0.2	0.1
Butyric Acid	mg/L	7	average	700	1300	1500	1500	1300
			st. dev.	200	200	100	200	200
			coeff. of var.	0.2	0.2	0.1	0.1	0.2
Isobutyric Acid	mg/L	7	average	400	1200	2200	1400	700
			st. dev.	20	1800	2700	1900	90
			coeff. of var.	0.1	1.5	1.2	1.3	0.1
Valeric Acid	mg/L	7	average	200	500	400	200	200
			st. dev.	30	70	40	20	20
			coeff. of var.	0.2	0.2	0.1	0.1	0.1
Isovaleric Acid	mg/L	7	average	200	500	700	800	800
			st. dev.	40	70	70	100	130
			coeff. of var.	0.2	0.1	0.1	0.1	0.1

Table 2-1. Process 1 Averages and Standard Deviations for Measured Parameters, continued.

Parameter	Units	# of samples		Feed	Meso 35	Thermo 48	Thermo 54	Thermo 62
Total Solids	percent	24	average	4.2	3.6	3.7	3.3	3.3
			st. dev.	0.7	0.4	1.2	0.3	0.4
			coeff. of var.	0.2	0.1	0.3	0.1	0.1
Volatile Solids	percent of g TS	24	average	81.3	78.2	78.2	77.3	77.2
			st. dev.	1.8	1.5	2.1	2.0	2.0
			coeff. of var.	0.02	0.02	0.03	0.03	0.03
VS Reduction	percent	26	average		19.0	19.1	22.1	21.5
			st. dev.		8.6	8.2	10.7	10.9
			coeff. of var.		0.5	0.4	0.5	0.5
Fecal Coliforms	MPN/g TS	24	average	300,000	15,000	5,000	1,000	400
			st. dev.	620,000	20,000	5,000	2,000	1,000
			coeff. of var.	2.1	1.3	1.0	2.0	2.5
Fecal Log reduction		24	average		1.2	1.7	3.0	3.6
			st. dev.		0.7	0.6	1.1	1.1
			coeff. of var.		0.5	0.4	0.4	0.3

Acid-phase digester temperatures were quite variable because of their low residence time and frequency of feed flows (Figures 2-3, 2-4, 2-5, and 2-6). Digester temperature standard deviation ranged from 2-3°C. Mean digester temperatures were 35, 48, 54, and 62°C. For discussion purposes in Section 2.3.1, the three thermophilic digesters will be referred to as Thermo 48, Thermo 54, and Thermo 62, respectively. Temperature fluctuations were mitigated by preheating the feed sludge and increasing the feed cycle to hourly (i.e., reducing the feed volume per cycle). Minimum digester temperatures usually reflected the colder feed sludge entering the digester because the temperature probe was located near the feed sludge inlet. This localized cold spot may exaggerate variation in digester temperature. Other minimum temperatures (e.g., Day 9) were the results of equipment malfunction or a temporary loss of heating capacity.

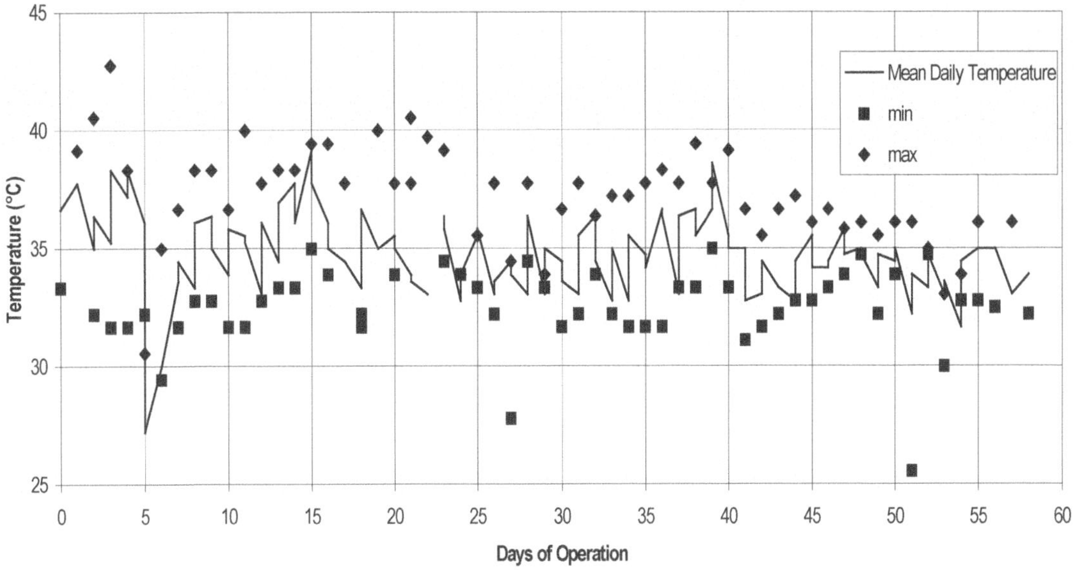

Figure 2-3. Process 1 Temperatures for Meso 35.

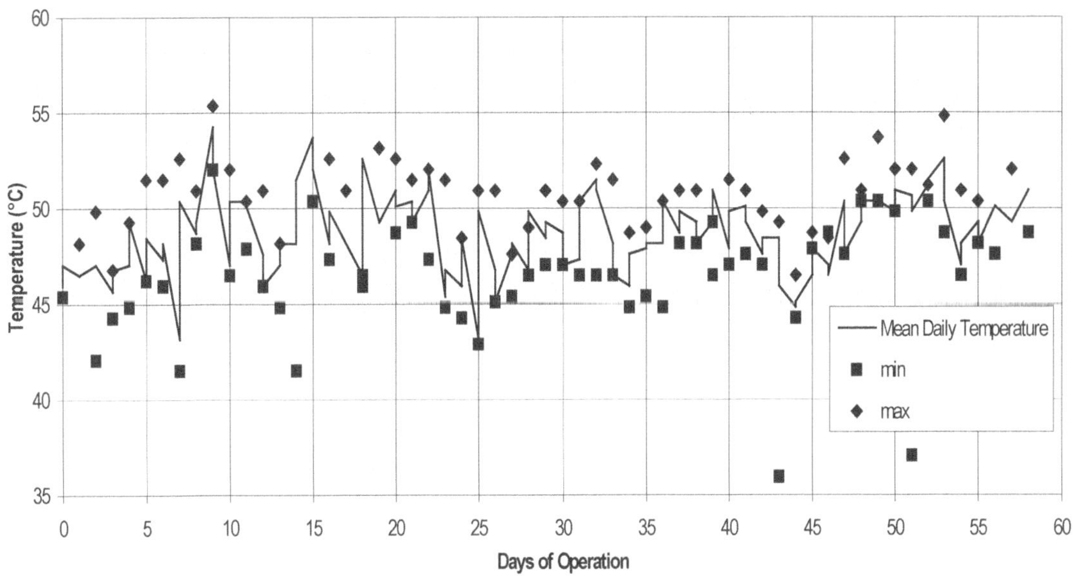

Figure 2-4. Process 1 Temperatures for Thermo 48.

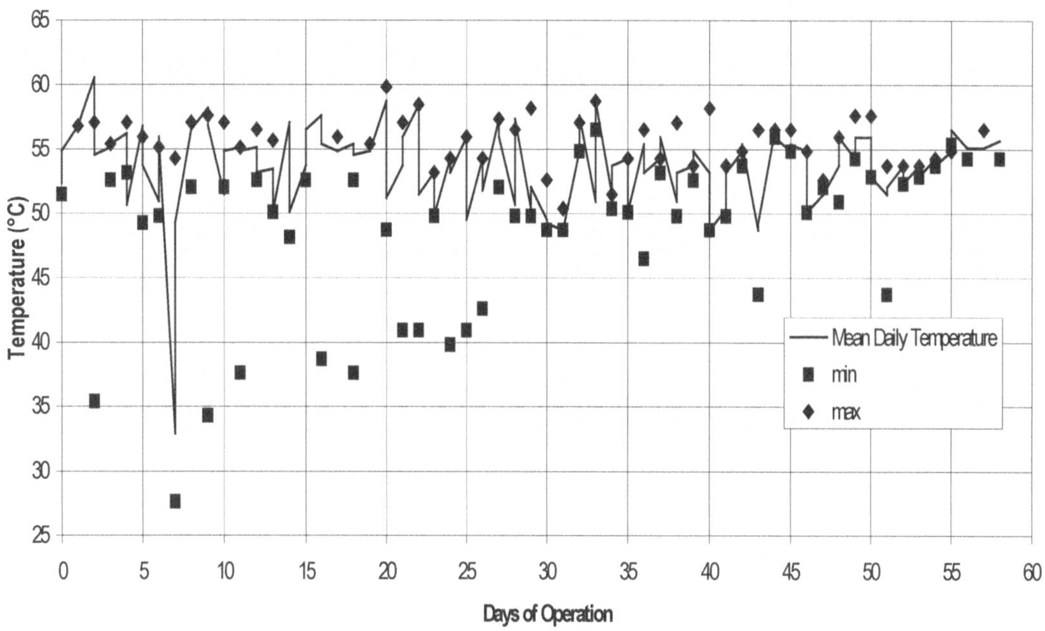

Figure 2-5. Process 1 Temperatures for Thermo 54.

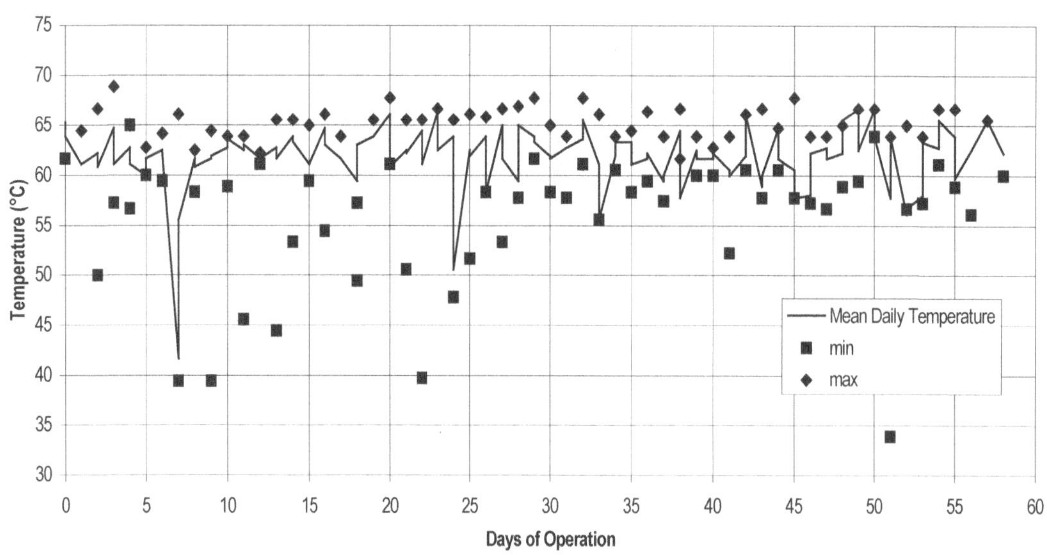

Figure 2-6. Process 1 Temperatures for Thermo 62.

The target hydraulic residence time (HRT) for all acid-phase digesters was 2 days. There was substantial variability in the HRT due to pumping issues and digester foaming (Figure 2-7). Average HRTs ranged from 2.5 to 3.4 days. Although HRTs were higher than the target value, they were still low enough to maintain consistently acid conditions.

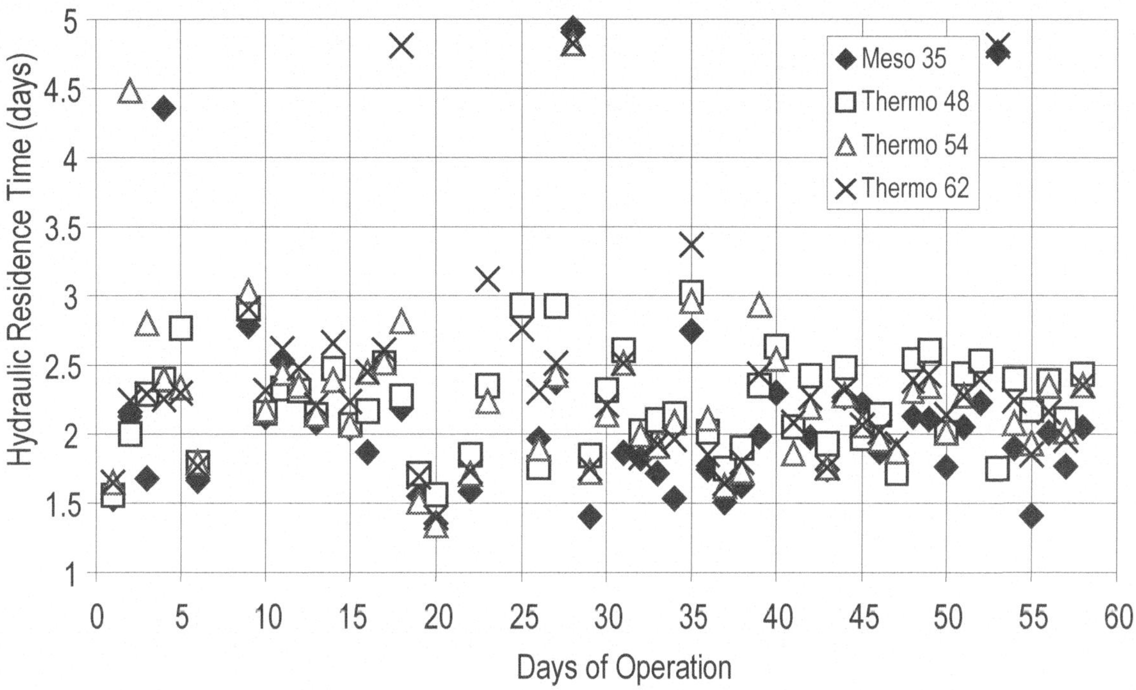

Figure 2-7. Process 1 Digester Hydraulic Residence Times.

Reactor pH and total alkalinity (TALK) appear to be related to the target operating temperature (Figures 2-8 and 2-9, respectively). The higher digester temperatures resulted in higher pH and TALK values. Average pH values ranged from 5.8 (Meso 35) to 6.6 (Thermo 62), and were positively correlated with temperature. Average TALK values ranged from approximately 3,400 mg/L $CaCO_3$ (Meso 35) to 5,100 mg/L $CaCO_3$ (Thermo 62). Total volatile acid (TVA) concentrations showed greater variability than TALK, and their relationship to temperature was not as strong (Figure 2-10). The average TVA concentrations for the Thermo digesters (ranging from 7,300 to 7,600 mg as acetic acid/L) were somewhat (but not significantly) higher than for the Meso 35 (approximately 6,700 mg/L). The TVA to TALK ratio ranged from an average of 1.5 (Thermo 62) to 2.0 (Meso 35) and was largely a function of TALK (Figure 2-11). A plot of mean specific volatile fatty acids (VFAs) for each digester shows that the VFA distribution did not vary significantly with digester temperature (Figure 2-12). The highest VFA concentrations were for acetic acid, followed by propionic, butyric, and isobutyric acids. The acid-phase digesters show significantly higher concentrations of the higher molecular weight VFAs than the single-stage or methane-phase digesters. (See discussion in Sections 2.3.2 and 2.3.3.)

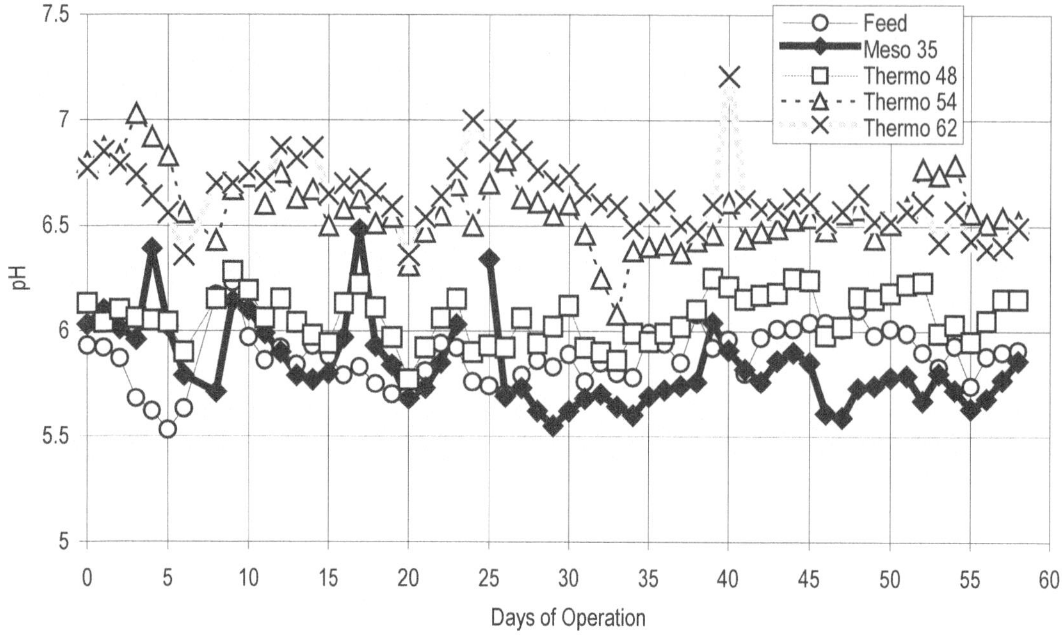

Figure 2-8. Process 1 pH.

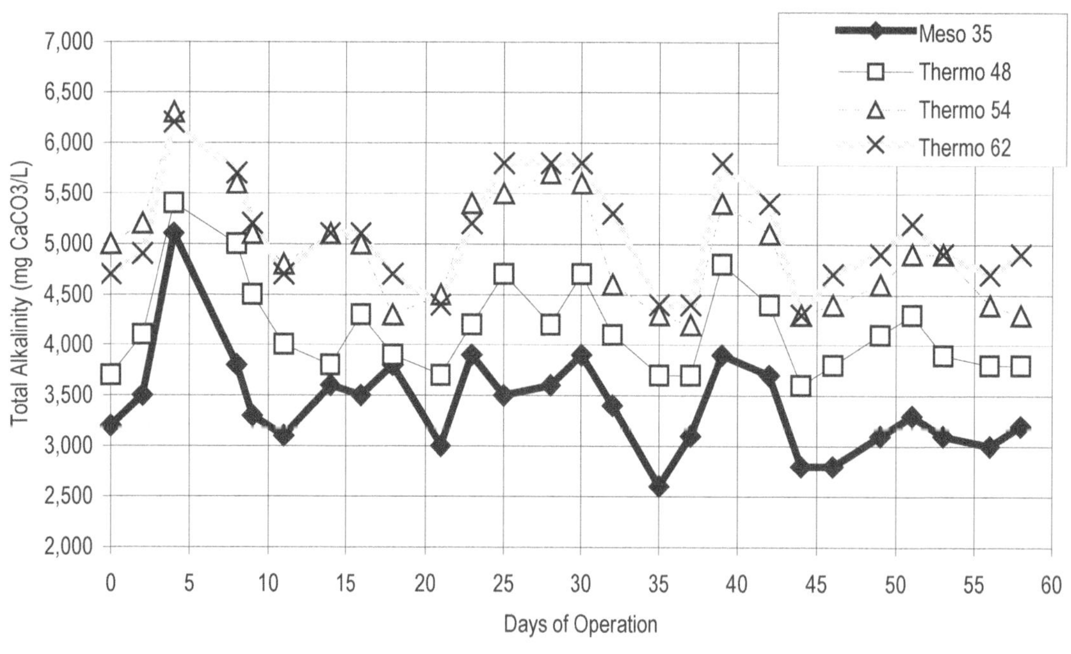

Figure 2-9. Process 1 Total Alkalinity.

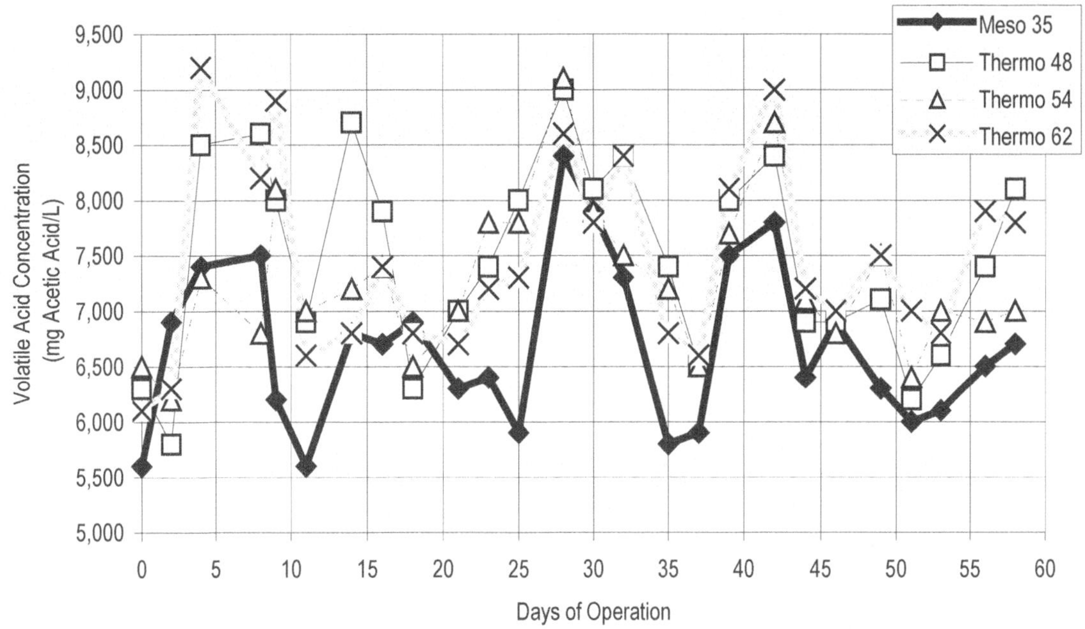

Figure 2-10. Process 1 Total Volatile Acid Concentration.

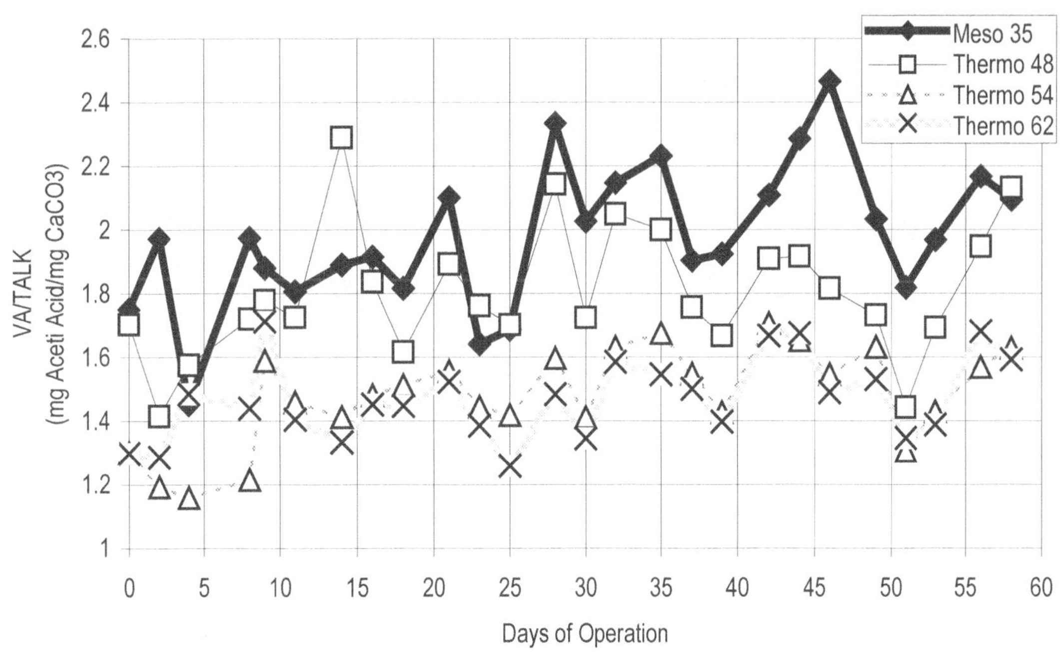

Figure 2-11. Process 1 Total Volatile Acid/Total Alkalinity Ratio (TVA/TALK).

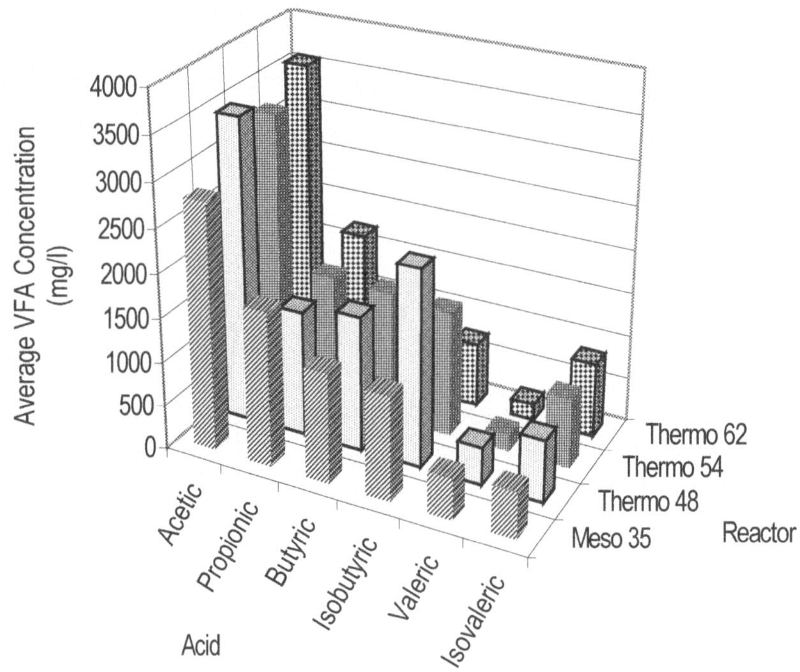

Figure 2-12. Process 1 Averages of Specific Volatile Fatty Acids.

Feed sludge volatile solids (VS) were generally between 80% and 85% percent of TS (Figure 2-13). Feed sludge percent VS was dependent upon thickening unit operation (gravity for PS and gravity belt thickener for TWAS). Average VS reduction (VSR) ranged from 19% (Meso 35) to 22% (Thermo 54), calculated with the Van Kleeck equation and a running average of influent VS. The running average was calculated over a period of 2 days (target HRT) prior to the collection date of the digested sludge sample. Average VSR appeared to increase slightly with temperature, (from 19% to 22%) as temperature increased from 48° to 54°C (Table 2-1).

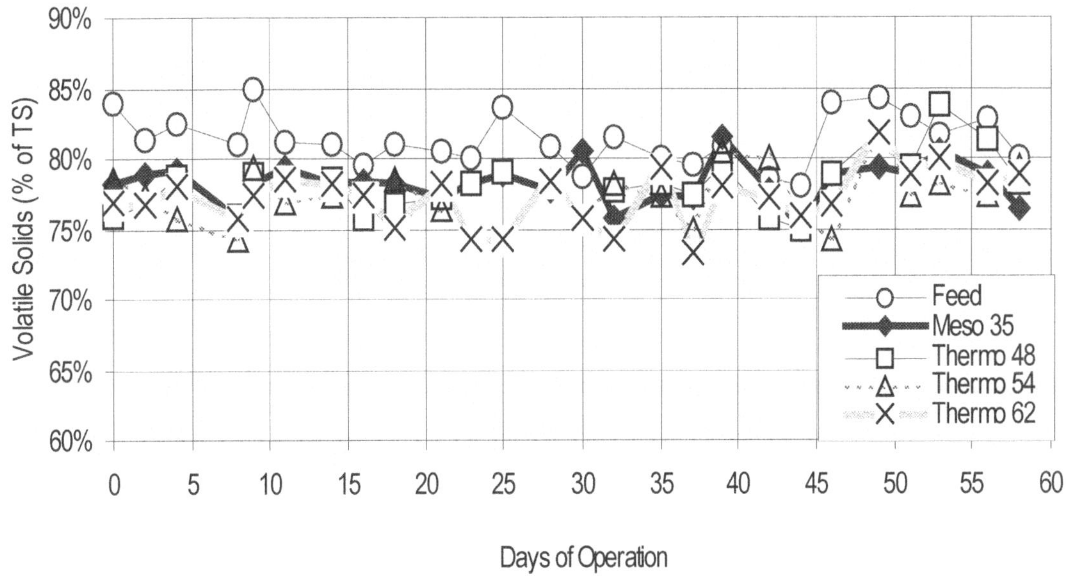

Figure 2-13. Process 1 Volatile Solids.

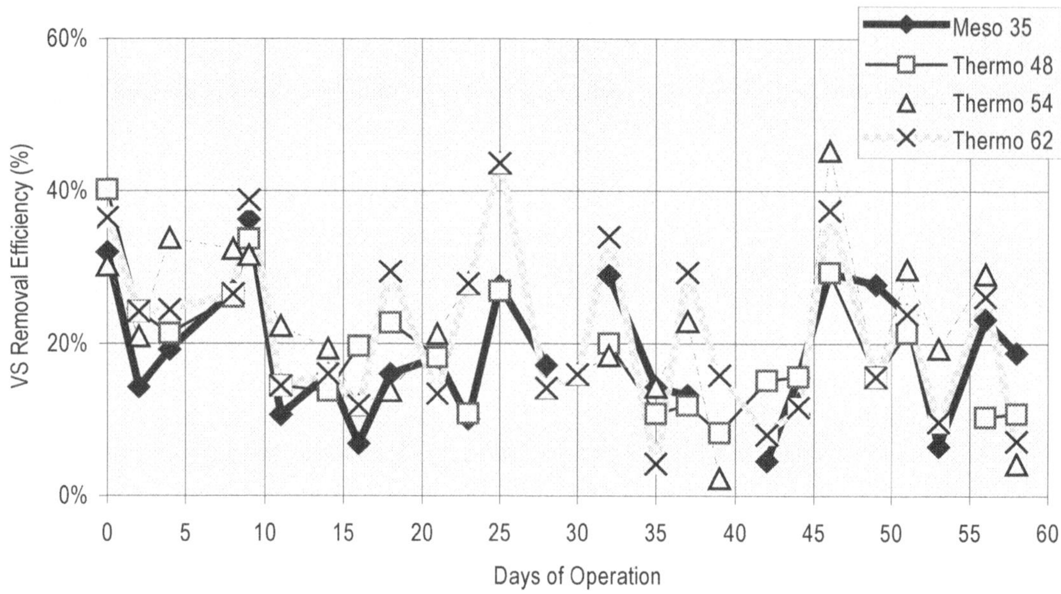

Figure 2-14. Process 1 Volatile Solids Reduction.

Fecal coliform concentrations and removals showed a consistent relationship to operating temperature (Figures 2-15 and 2-16). Mean fecal coliform reductions for the 54° and 62° C reactors were substantially higher than those for the 35° and 48° C reactors. However, none of the digesters consistently achieved the Class A criterion despite the relatively low fecal coliform densities in the feed sludge (approximately 10^5 MPN/g TS). A comparison of fecal coliform destruction for the three processes at different temperatures is discussed in detail in Section 2.3.5.

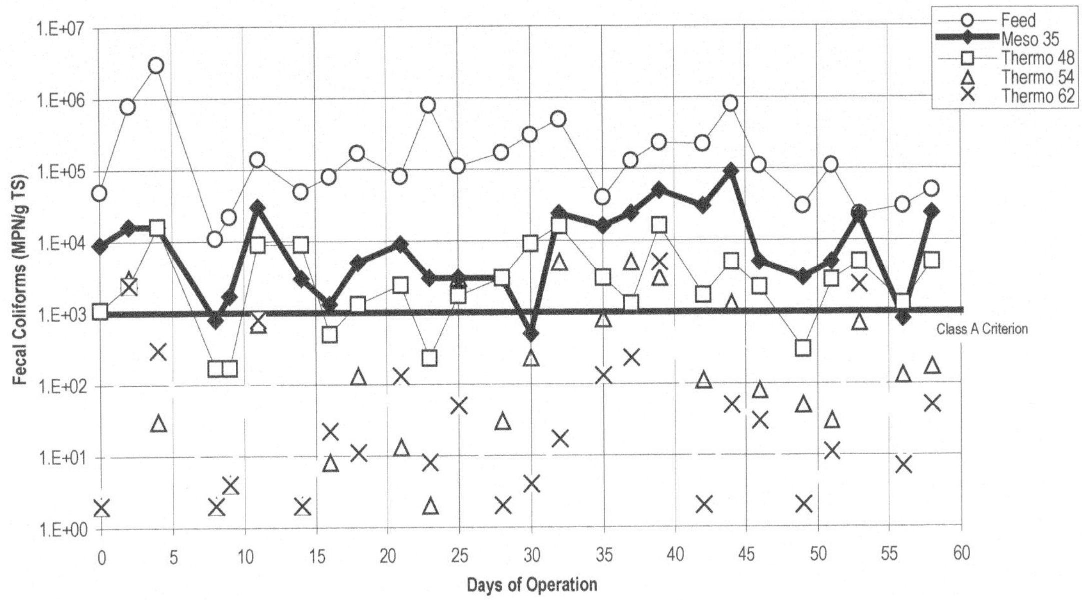

Figure 2-15. Process 1 Fecal Coliforms.

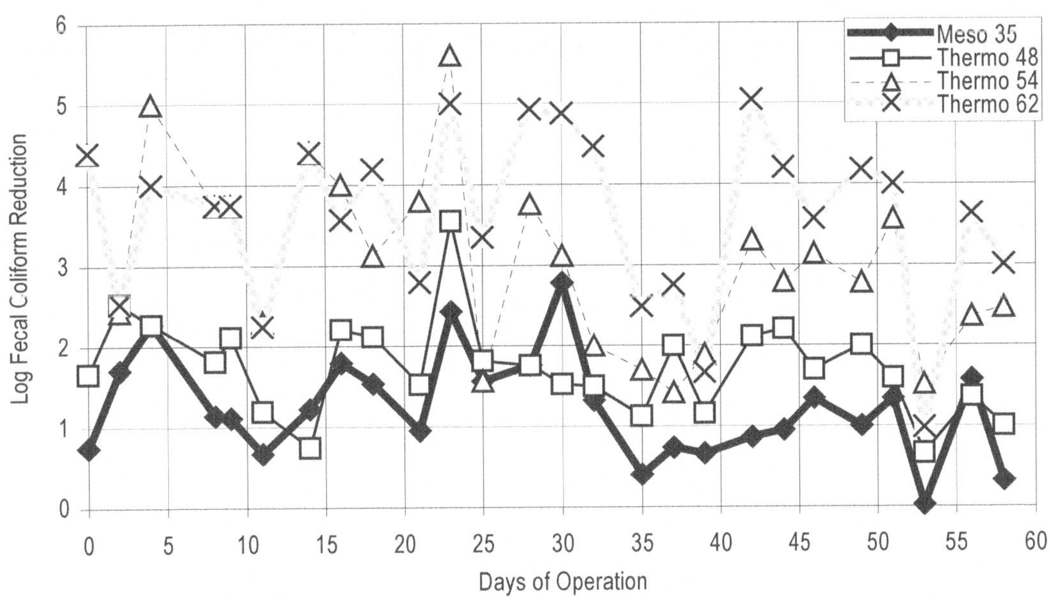

Figure 2-16. Process 1 Fecal Coliform Log Reduction.

2.3.2 Process 2: Thermophilic Single Stage

Single-stage thermophilic digestion was the second process in the screening evaluation. Three single-stage thermophilic digesters were operated at average temperatures of 49, 53, and 62° C (Figures 2-17). For discussion purposes in Section 2.3.2, the three thermophilic digesters will be referred to as Thermo 49, Thermo 53, and Thermo 62, respectively. The mesophilic control digester is referred to as Meso 35. Target HRT for all digesters was 15 days. Table 2-2 lists average values, standard deviations, and coefficients of variation for parameters measured for Process 2.

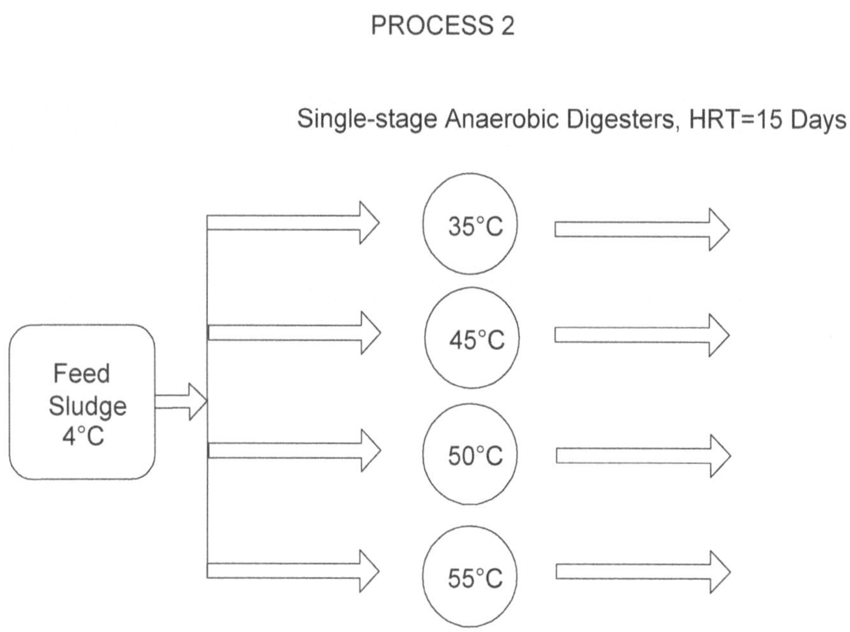

Figure 2-17. Phase 1 Process 2 Schematic.

Table 2-2. Process 2 Averages and Standard Deviations for Measured Parameters.

Parameter	Units	# of samples		Feed	Meso 35	Thermo 49	Thermo 53	Thermo 62
Residence time	days	86	average		15	15	13	15
			st. dev.		4	3	3	4
			coeff. of var		0.2	0.2	0.2	0.3
Temperature	°C	167	average		34	49	53	62
			st. dev.		2	2	3	2
			coeff. of var		0.06	0.04	0.05	0.03
Gas production	L/hr	86	average		1.6	1.7	1.6	1.2
			st. dev.		0.6	1.0	0.6	0.9
			coeff. of var		0.3	0.6	0.4	0.7

WERF

Table 2-2. Process 2 Averages and Standard Deviations for Measured Parameters, continued.

Parameter	Units	# of samples		Feed	Meso 35	Thermo 49	Thermo 53	Thermo 62
Methane in gas produced	percent	16	average		63	59	56	55
			st. dev.		7.3	10.6	10.9	9.1
			coeff. of var		0.12	0.18	0.19	0.17
pH		85	average	5.9	7.9	8.0	8.1	8.0
			st. dev.	0.2	0.1	0.2	0.1	0.2
			coeff. of var		0.0	0.0	0.0	0.0
Total Alkalinity	mg $CaCO_3$/L	46	average		6800	6900	6700	6400
			st. dev.		500	500	400	600
			coeff. of var		0.1	0.1	0.1	0.1
Total Volatile Acids	mg acetic acid/L	46	average		300	300	1600	2400
			st. dev.		300	200	800	1200
			coeff. of var		1.0	0.7	0.5	0.5
TVA/TALK ratio	mg acetic acid/mg $CaCO_3$	46	average		0.04	0.05	0.2	0.4
			st. dev.		0.05	0.03	0.1	0.2
			coeff. of var		1.0	0.7	0.5	0.6
Acetic Acid	mg/L	45	average	1600	300	400	500	1000
			st. dev.	700	200	100	200	900
			coeff. of var	0.4	0.7	0.3	0.4	0.9
Propionic Acid	mg/L	45	average	800	70	80	800	1000
			st. dev.	400	100	100	500	400
			coeff. of var	0.5	1.4	1.3	0.6	0.4
Butyric Acid	mg/L	45	average	600	30	30	40	100
			st. dev.	400	20	4	40	200
			coeff. of var	0.7	0.7	0.1	1.0	2.0
Isobutyric Acid	mg/L	45	average	300	70	90	160	300
			st. dev.	100	30	30	70	200
			coeff. of var	0.3	0.4	0.3	0.4	0.7
Valeric Acid	mg/L	45	average	90	26	25	40	40
			st. dev.	70	4	1	30	20
			coeff. of var	0.8	0.2	0.1	0.8	0.5
Isovaleric Acid	mg/L	45	average	200	30	30	60	200
			st. dev.	100	20	10	70	200
			coeff. of var	0.5	0.7	0.3	1.2	1.0
Total Solids	percent	46	average	4.5	2.3	2.3	2.3	2.3
			st. dev.	1.1	0.3	0.3	0.3	0.3
			coeff. of var	0.2	0.1	0.1	0.1	0.1
Volatile Solids	percent of TS	46	average	76.9	64.7	64.1	65.9	66.3
			st. dev.	4.6	3.2	4.1	2.6	3.5
			coeff. of var	0.06	0.05	0.06	0.04	0.05
VS Reduction	percent	46	average		48.9	50.8	45.6	44.3
			st. dev.		5.8	6.4	5.6	6.9
			coeff. of var		0.1	0.1	0.1	0.2

Table 2-2. Process 2 Averages and Standard Deviations for Measured Parameters, continued.

Parameter	Units	# of samples		Feed	Meso 35	Thermo 49	Thermo 53	Thermo 62
Fecal Coliforms	MPN/g TS	46	average	160,000	1,700	300	100	20
			st. dev.	170,000	1,900	700	300	80
			coeff. of var	1.1	1.1	2.3	3.0	4.0
Fecal Log reduction		46	average		2.0	3.0	3.8	4.3
			st. dev.		0.5	0.7	0.8	0.7
			coeff. of var		0.3	0.2	0.2	0.2

Process 2 digester temperatures were generally less variable than those for Process 1 because smaller flows of cold feed sludge were used at the longer HRT (Figures 2-18 through 2-21). The preheating system used for Process 1 (acid-phase) was not required for Process 2. However, Thermo 49 and Thermo 53 temperatures fluctuated significantly (>10° C) on several occasions because of problems with the temperature controller. These isolated events are responsible for most of the observed temperature variability of these two digesters.

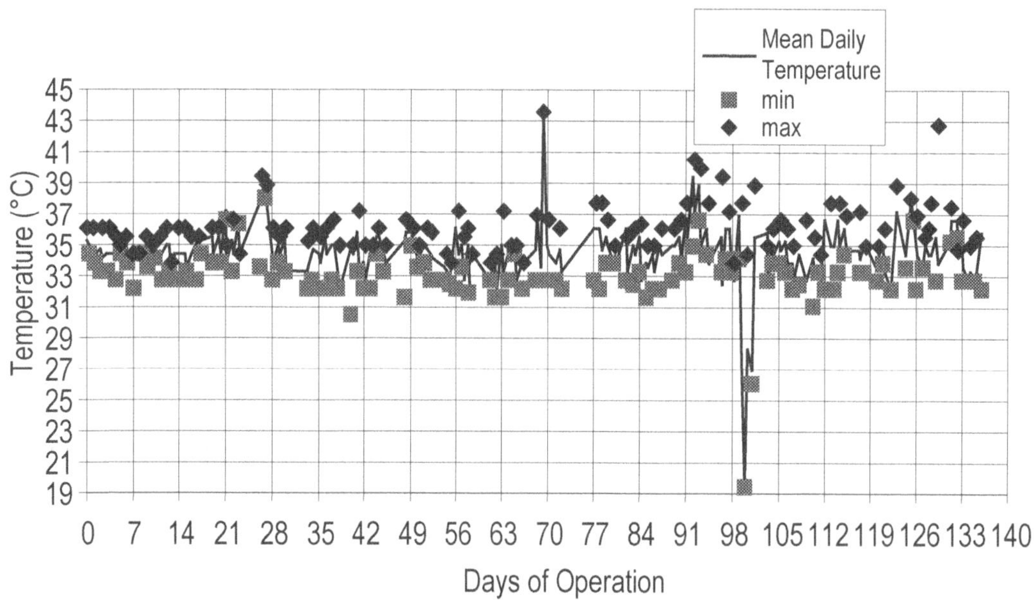

Figure 2-18. Process 2 Temperatures for Meso 35.

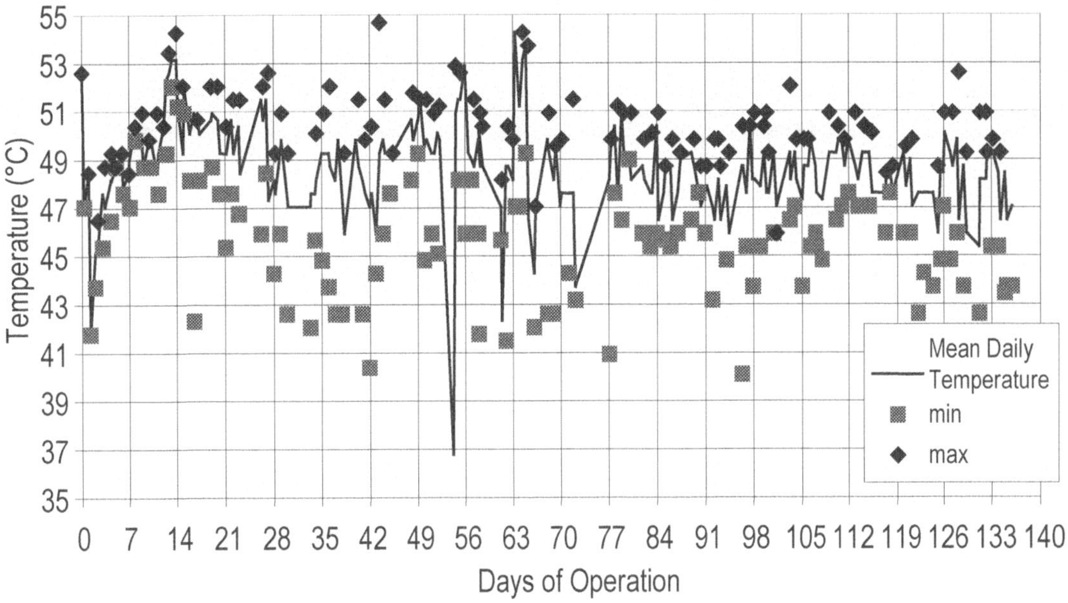

Figure 2-19. Process 2 Temperatures for Thermo 49.

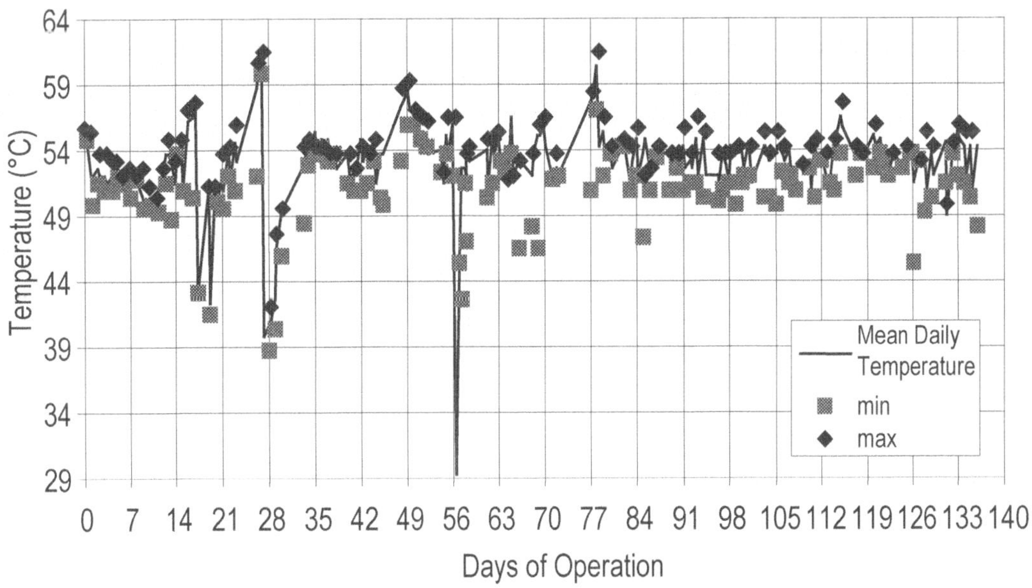

Figure 2-20. Process 2 Temperatures for Thermo 53.

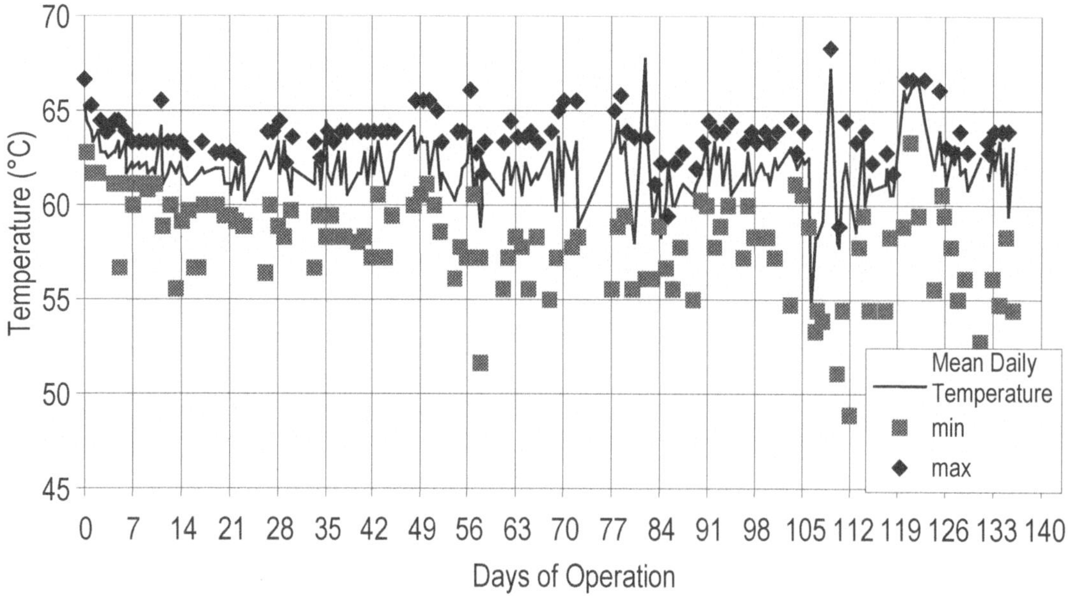

Figure 2-21. Process 2 Temperatures for Thermo 62.

To allow for a more gradual start-up period, the digesters were operated at an HRT higher than the 15-day target (Figure 2-22). The pumping rate and influent/effluent flows were increased after about 2 weeks to achieve the target 15-day HRT.

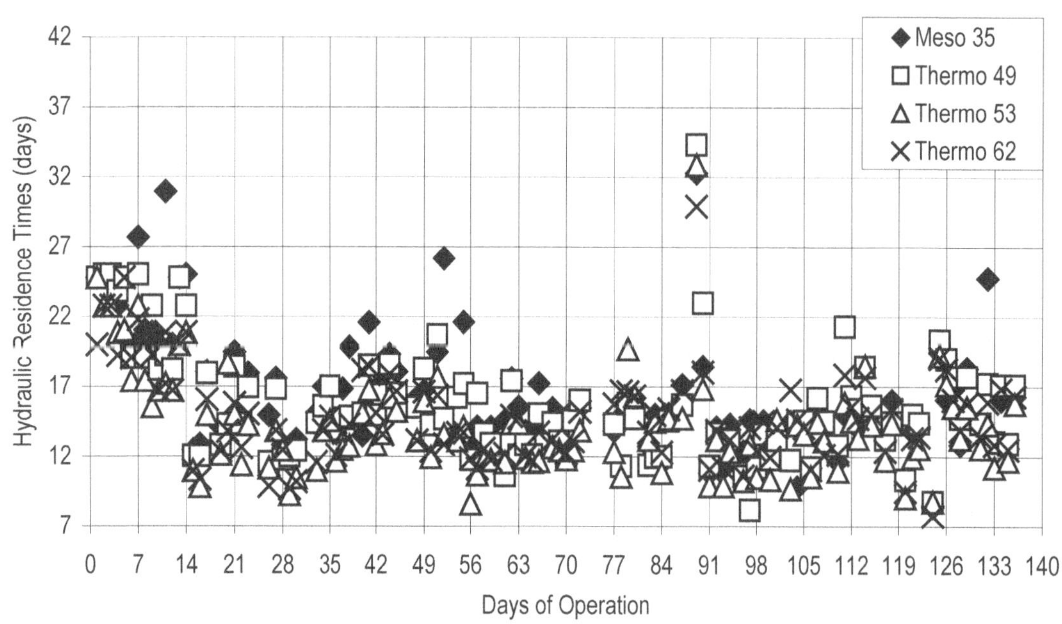

Figure 2-22. Process 2 Digester Hydraulic Residence Times.

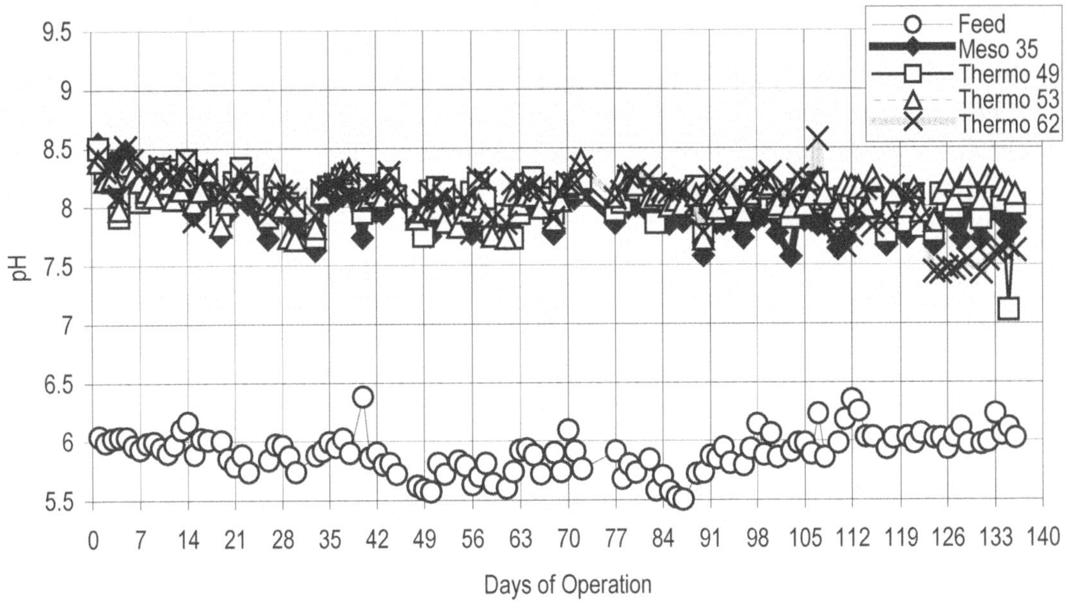

Figure 2-23. Process 2 pH.

The feed pH was approximately 6, and the pH in all 4 digesters remained around 8 (Figure 2-23). This is a relatively high pH for anaerobic digestion, but the high TALK and the low TVA concentration supported these results. Digester TALKs were relatively high with an average concentration of approximately 6,600 mg $CaCO_3$/L (Figure 2-24). There are no significant differences in TALK among the digesters.

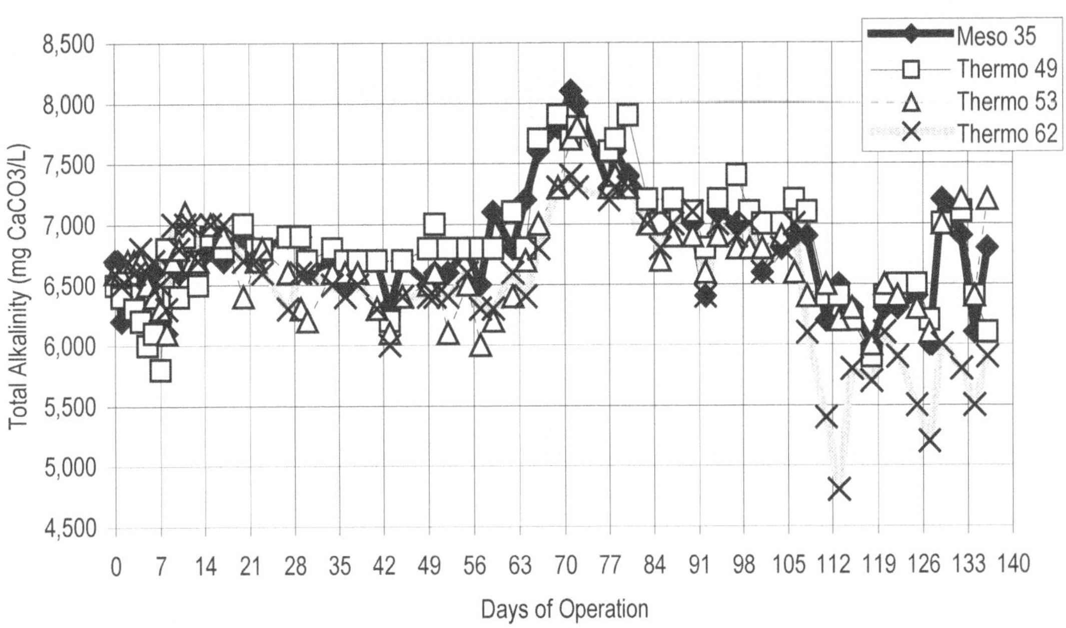

Figure 2-24. Process 2 Total Alkalinity.

The TVA levels were higher for the Thermo 53 and 62 reactors than for the Meso 35 and the Thermo 49 digesters (Figure 2-25). The low TVA/TALK ratios measured in the Meso 35 and Thermo 49 digesters were similar to those of stable digesters, i.e., less than 0.10 (Figure 2-26). The TVA/TALK ratios measured in the Thermo 53 and Thermo 62 digesters were significantly higher (approximately 0.2 to 0.4). These two digesters also had the highest propionic acid concentrations (Figure 2-27). Figure 2-28 further suggests that there is a strong correlation between temperature and the amount of acetic, propionic, isobutyric, valeric, and isovaleric acid in digested sludge. Because these VFAs are odorous, operating thermophilic digesters at lower temperature may reduce the odors that have been associated with thermophilic anaerobic digestion. If this is the case, then the Thermo 49 digester would produce biosolids no more odorous than a mesophilic digester would.

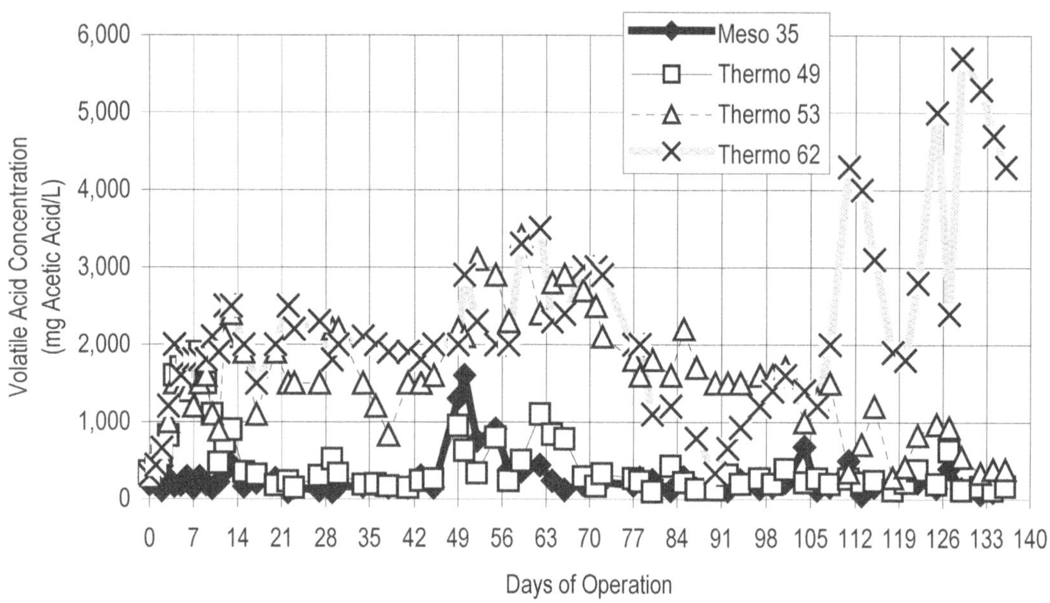

Figure 2-25. Process 2 Total Volatile Acid Concentration.

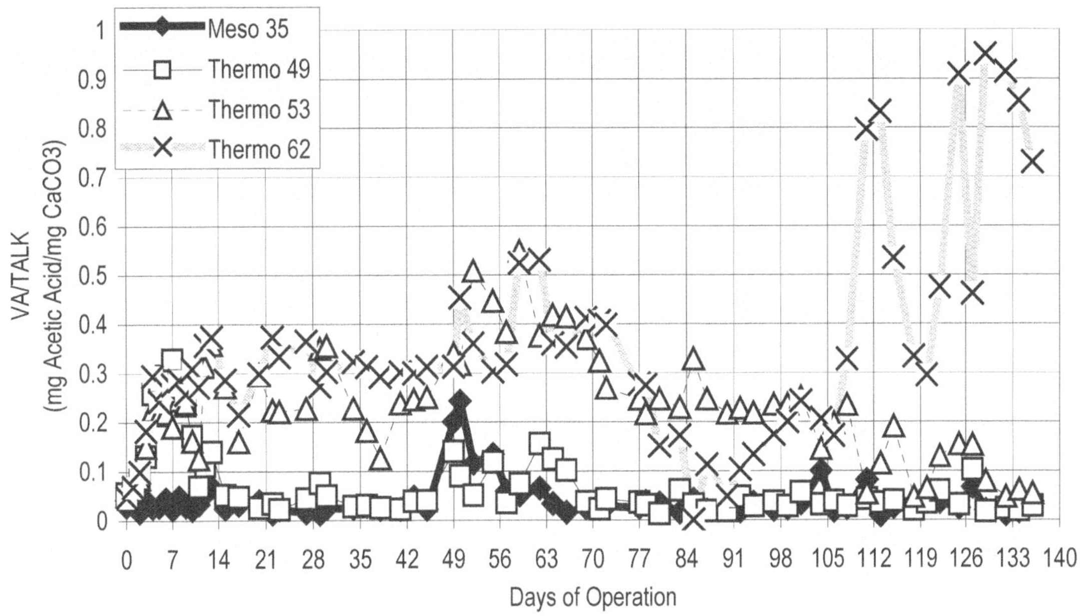

Figure 2-26. Process 2 Total Volatile Acid/Total Alkalinity Ratio (TVA/TALK).

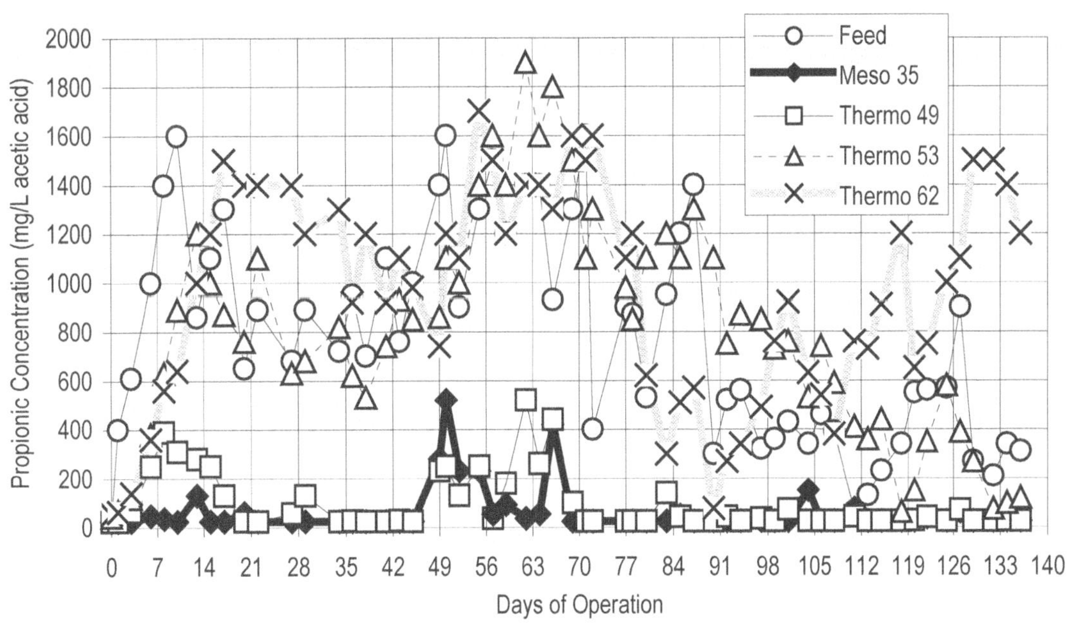

Figure 2-27. Process 2 Propionic Acid Concentration.

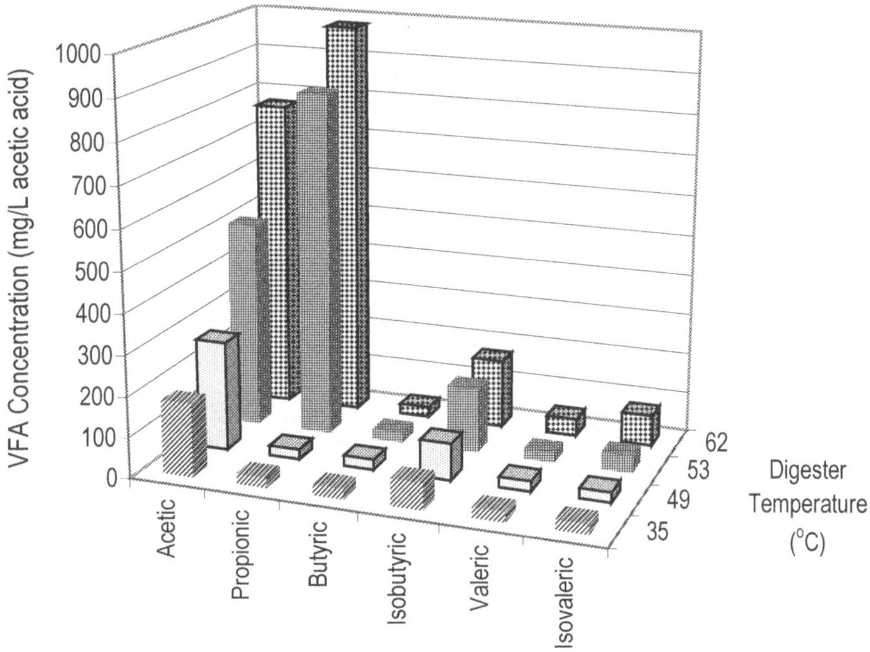

Figure 2-28. Process 2 Specific Volatile Fatty Acid Concentrations at Process Equilibrium.

Generally, VS concentrations were similar for all digesters in the range of 60% to 75% (Figure 2-29). The VSR is generally above 40% for all digesters (Figure 2-30). VSRs were calculated using the Van Kleeck equation.

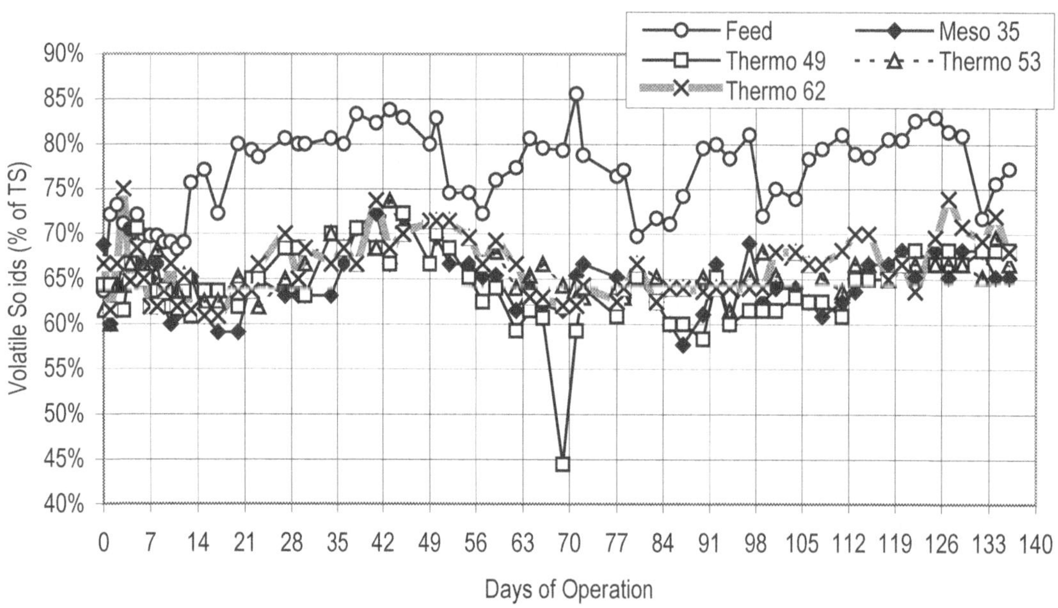

Figure 2-29. Process 2 Volatile Solids.

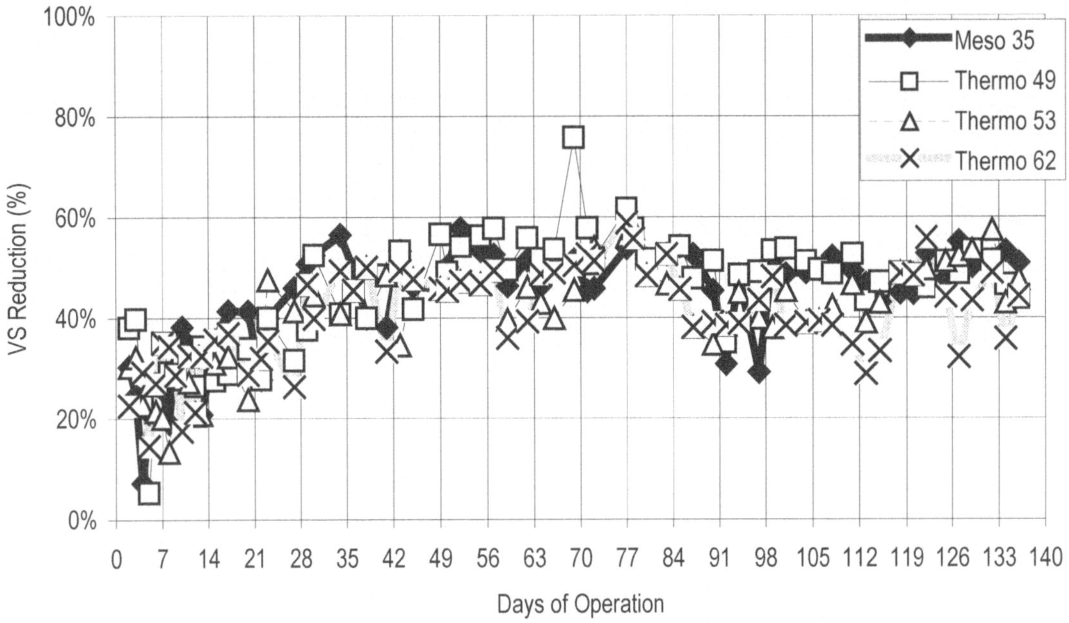

Figure 2-30. Process 2 Volatile Solids Reduction.

The arithmetic average fecal coliform concentration in the digested sludge shows an inverse relationship with respect to temperature ranging from 1,700 MPN/g TS for Meso 35 to 20 MPN/g TS for Thermo 62. All of the digesters generally met the Class A fecal coliform criterion of <1,000 MPN/g TS, except Meso 35 (Figure 2-31). Feed fecal coliform concentrations were relatively low (between 10^4 and 10^5), which probably explains why the Meso digester was able at times to meet the criterion.

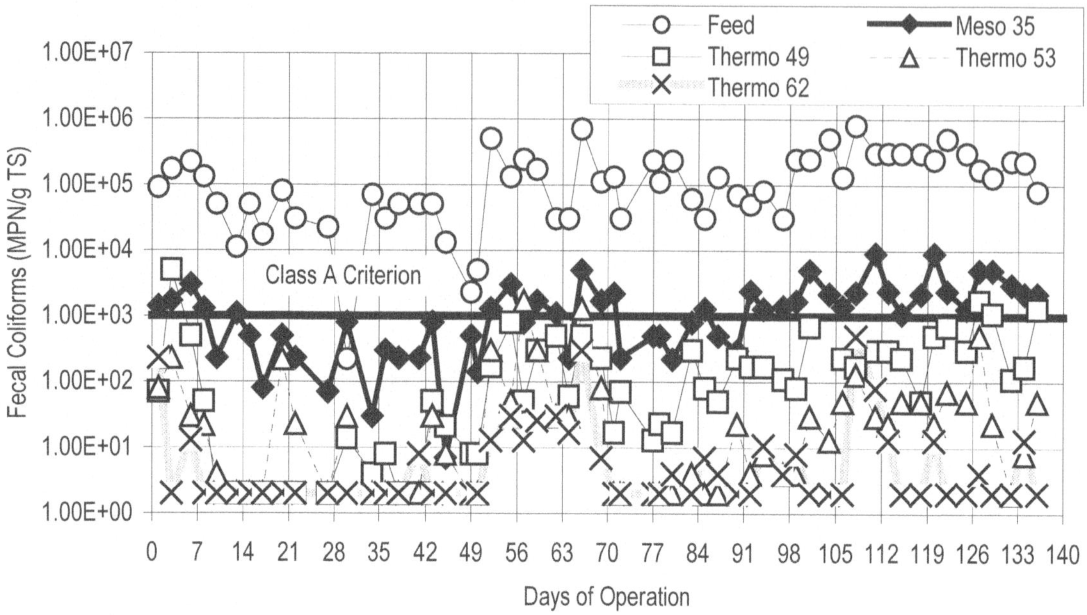

Figure 2-31. Process 2 Fecal Coliforms.

The literature states that fecal coliform log reductions in thermophilic anaerobic digesters have ranged between 4 and 8 (Aitken and Mullenix, 1992; Berg and Berman, 1980; Garber, 1982, 1977; Lee et al., 1989; Huyard et al., 1998). The fecal coliform log reductions observed in the Process 2 thermophilic digesters are near the low end of this range (Figure 2-32). It is possible that these low values were caused by digester temperature fluctuations. It is also possible that the lower feed concentrations decrease log reduction potential. Low fecal coliform concentrations in the EBMUD feed sludge could be the result of prechlorination for odor control at the plant headworks.

Comparing Figures 2-31 and 2-26, for data between Days 110 and 135, it is possible to conclude that thermophilic fecal coliform destruction is independent of process stability. Anaerobic digestion is usually considered stable at TVA/TALK ratio of <0.10. Between Days 110 and 135, Thermo 62 VA/TALK fluctuated between 0.30 to >0.90, suggesting an unstable process. During this same period, the Thermo 62 fecal coliform densities were all below the Class A standard.

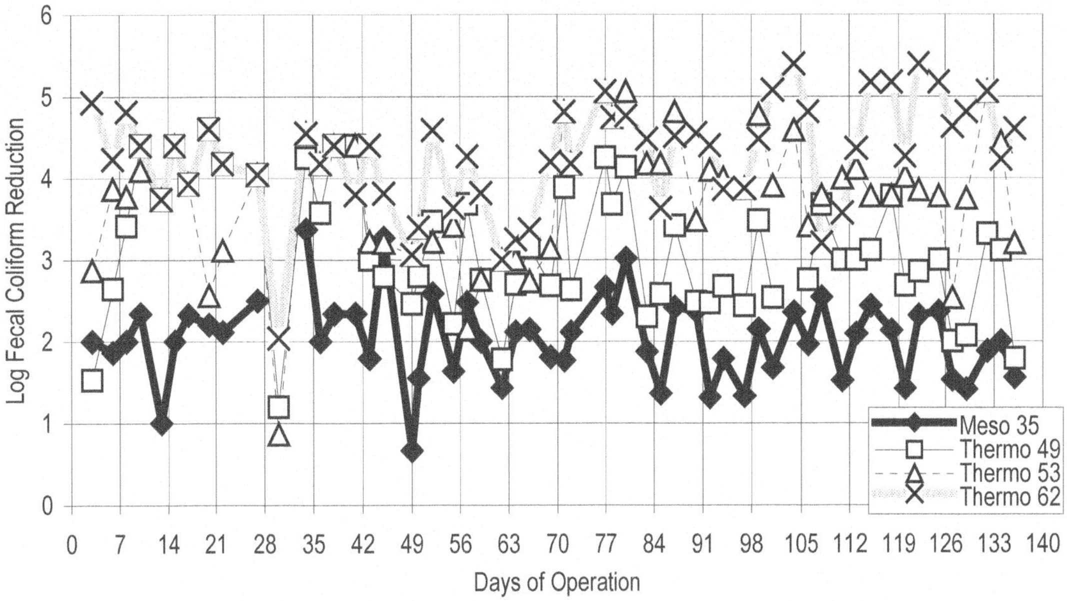

Figure 2-32. Process 2 Fecal Coliform Log Reduction.

2.3.3 Process 3: Two-Stage Meso Acid/Thermo Methane

Two-stage digestion using a mesophilic acid phase followed by a thermophilic methane phase (Meso acid/Thermo methane) was the third process tested as part of the screening evaluation. A single mesophilic acid-phase digester fed four methane phase digesters: a meso control and three thermophilic digesters operated at average temperatures of 48, 52, and 62° C (Figure 2-33). For discussion purposes in Section 2.3.3, the three thermophilic digesters will be referred to as Thermo 48, Thermo 52, and Thermo 62, respectively. The mesophilic acid-phase digester is referred to as Meso Acid and the control methane-phase digester is referred to as Meso 35. Target HRTs for the acid-phase and methane-phase digesters were 2 and 15 days, respectively. Table 2-3 lists average values, standard deviations, and coefficients of variation for parameters measured for Process 3.

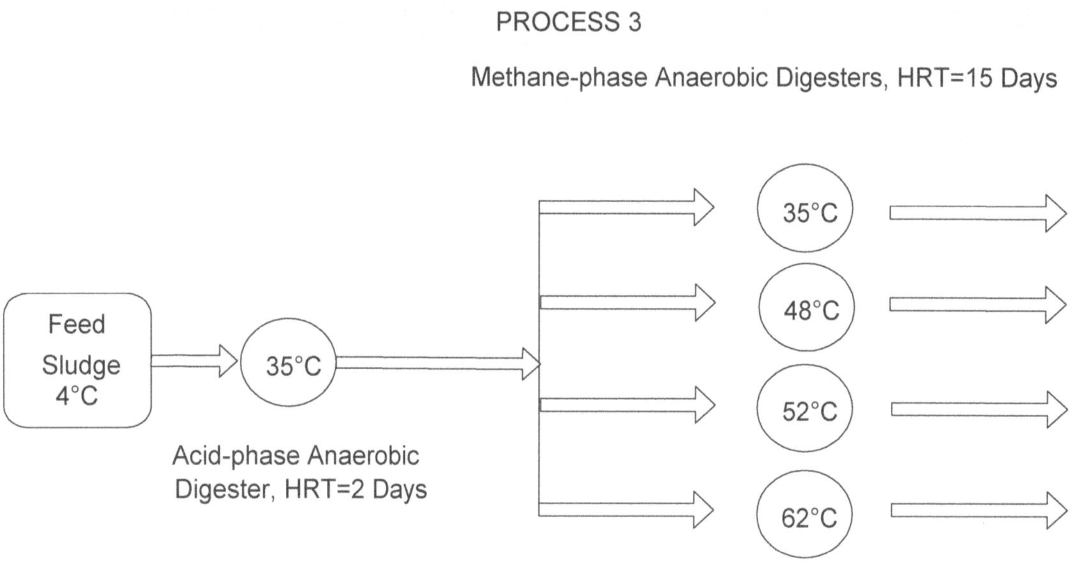

Figure 2-33. Phase1 Process 3 Schematic.

Table 2-3. Process 3 Averages and Standard Deviations for Measured Parameters.

Parameter		# of samples		Feed	Meso Acid	Meso 35	Thermo 48	Thermo 52	Thermo 62
Residence time	days	64	average		2	14	14	14	16
			st. dev.		0.5	2	2	2	2
			coeff. of var		0.3	0.1	0.1	0.1	0.2
Temperature	°C	116	average		36	34	48	52	62
			st. dev.		1	1	2	2	1
			coeff. of var		0.03	0.03	0.04	0.04	0.02
Gas production	L/hr	64	average		0.8	1.5	1.6	1.0	0.4
			st. dev.		0.5	0.6	0.6	0.8	0.3
			coeff. of var		0.61	0.39	0.42	0.74	0.77

Table 2-3. Process 3 Averages and Standard Deviations for Measured Parameters, continued.

Parameter		# of samples		Feed	Meso Acid	Meso 35	Thermo 48	Thermo 52	Thermo 62
Methane in gas produced	percent	11	average		24	68	67	55	49
			st. dev.		6	1	2	10	10
			coeff. of var		0.3	0.02	0.02	0.2	0.2
pH		64	average	6.4	5.8	8.2	8.3	7.7	7.2
			st. dev.	0.3	0.2	0.1	0.1	0.3	0.1
			coeff. of var	0.05	0.03	0.01	0.01	0.04	0.01
Total Alkalinity	mg $CaCO_3$/L	33	average		2800	6500	6700	5500	4600
			st. dev.		600	300	600	900	600
			coeff. of var		0.2	0.05	0.09	0.2	0.1
Total Volatile Acids	mg acetic acid/L	33	average		6500	100	600	5900	7300
			st. dev.		1000	60	400	1000	900
			coeff. of var		0.2	0.6	0.7	0.2	0.1
TVA/TALK ratio	mg acetic acid/mg $CaCO_3$	54	average		2.3	0.02	0.1	1.1	1.6
			st. dev.		0.5	0.01	0.1	0.3	0.1
			coeff. of var		0.2	0.6	0.7	0.2	0.1
Acetic Acid	mg/L	22	average	1100	2600	200	400	1500	3000
			st. dev.	200	400	40	90	600	400
			coeff. of var	0.2	0.2	0.2	0.2	0.4	0.1
Propionic Acid	mg/L	22	average	500	1500	30	200	1900	1800
			st. dev.	100	400	10	300	200	300
			coeff. of var	0.2	0.3	0.3	1.5	0.1	0.2
Butyric Acid	mg/L	22	average	300	800	25	25	600	800
			st. dev.	100	200	0	0	300	100
			coeff. of var	0.3	0.3	0	0	0.5	0.1
Isobutyric Acid	mg/L	22	average	100	300	30	60	500	500
			st. dev.	40	100	5	10	100	70
			coeff. of var	0.4	0.3	0.2	0.2	0.2	0.1
Valeric Acid	mg/L	22	average	60	400	25	25	300	300
			st. dev.	20	80	0	0	80	0
			coeff. of var	0.3	0.2	0	0	0.3	0.0
Isovaleric Acid	mg/L	22	average	100	400	25	30	800	800
			st. dev.	40	80	0	4	200	100
			coeff. of var	0.4	0.2	0	0.1	0.3	0.1
Total Solids	percent	33	average	4.3	3.6	2.1	2.0	2.4	2.2
			st. dev.	1.0	0.9	0.4	0.2	0.3	0.6
			coeff. of var	0.2	0.3	0.2	0.1	0.1	0.3
Volatile Solids	percent of TS	33	average	83	80	69	68	73	77
			st. dev.	2.7	3.4	4.0	3.0	2.4	2.6
			coeff. of var	0.03	0.04	0.06	0.04	0.03	0.03
VS Reduction	percent	33	average		18	53	55	41	30
			st. dev.		11	9	6	11	12
			coeff. of var		0.6	0.2	0.1	0.3	0.4

Table 2-3. Process 3 Averages and Standard Deviations for Measured Parameters, continued.

Parameter		# of samples		Feed	Meso Acid	Meso 35	Thermo 48	Thermo 52	Thermo 62
Fecal Coliforms	MPN/g TS	32	average	300,000	11,000	200	4	2	2
			st. dev.	860,000	11,000	200	5	0	0
			coeff. of var	2.9	1.0	1.00	1.1	0.2	0.0
Fecal Log reduction		32	average		1.2	3.1	4.6	4.7	4.8
			st. dev.		0.5	0.6	0.7	0.5	0.5
			coeff. of var		0.5	0.2	0.1	0.1	0.1

Average digester HRTs were close to the target values—2 days for the Meso Acid and from 14 to 16 days for the methane-phase reactors. After some early variability, the HRTs for all digesters were relatively close to the target values (Figure 2-34).

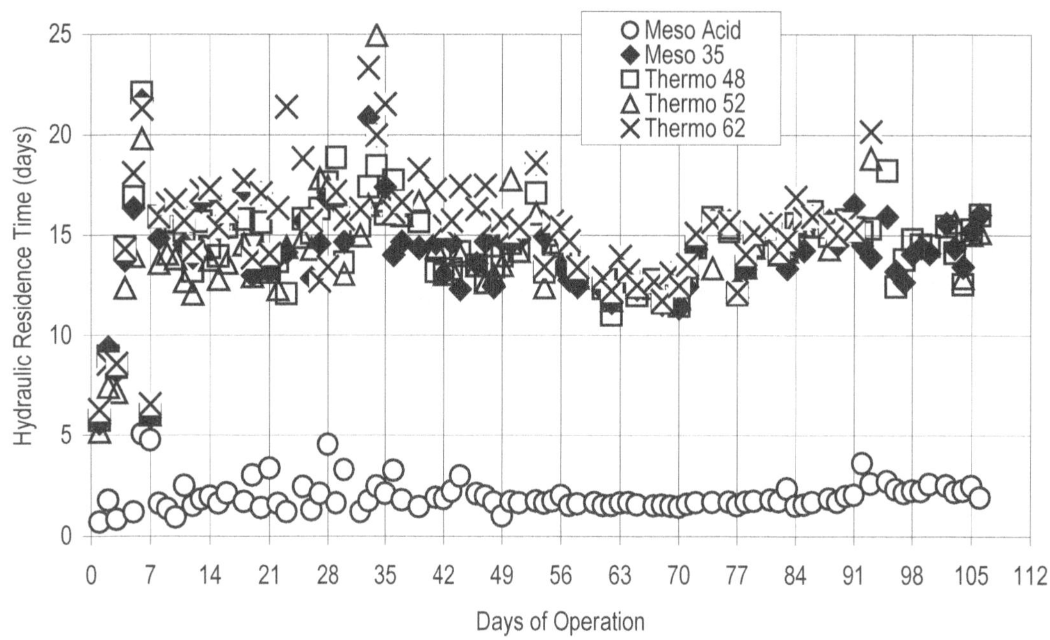

Figure 2-34. Process 3 Digester Hydraulic Residence Times.

Digester temperature data for Meso Acid, Meso 35, Thermo 48, Thermo 52, and Thermo 62 are in Figures 2-35, 2-36, 2-37, 2-38, and 2-39, respectively. Temperature variation was greatest in the Meso Acid, Thermo 48, and Thermo 62 digesters. Temperature variation in the Meso Acid was largely due to the higher influent flows of cold feed sludge. Temperature variation in Thermo 48 and Thermo 62 was due to difficulties with temperature controller operation. This variation sporadically could have been as great as 10° C in these two methane-phase digesters and undoubtedly contributed to process instability.

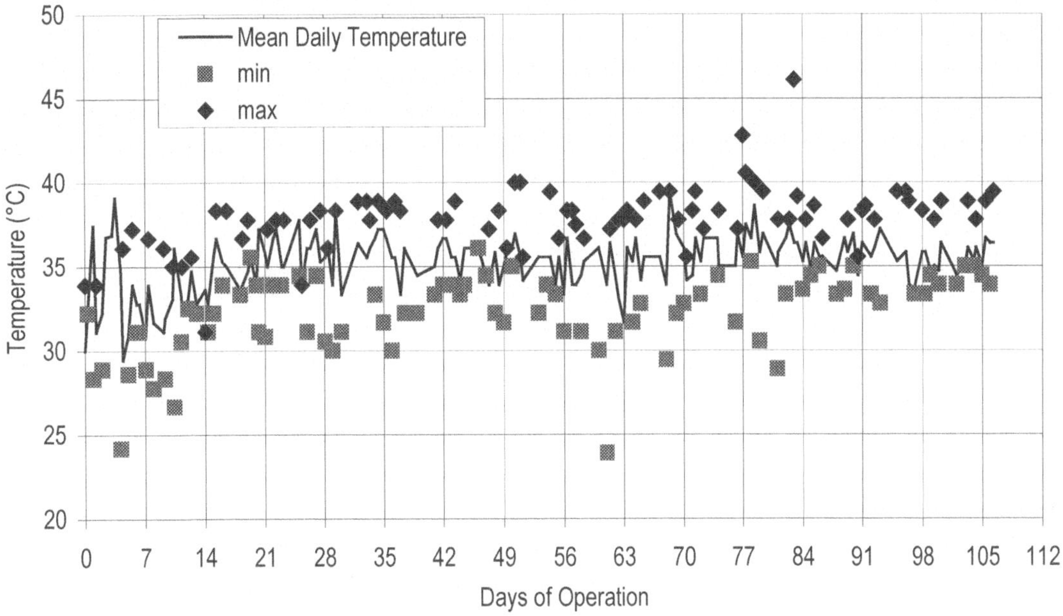

Figure 2-35. Process 3 Temperatures for Meso Acid.

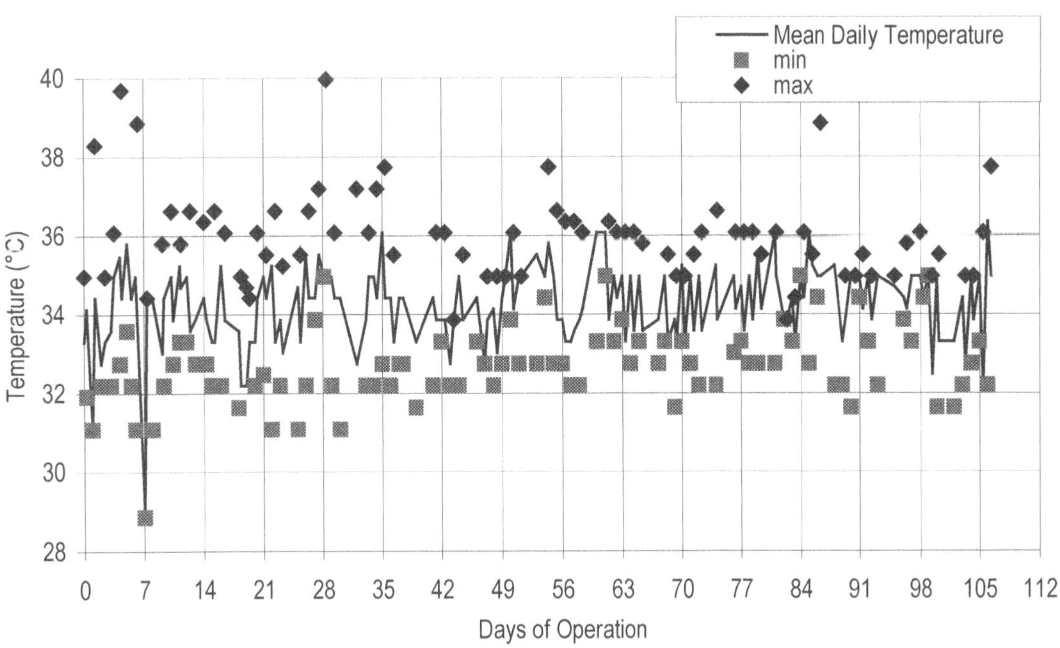

Figure 2-36. Process 3 Temperatures for Meso 35.

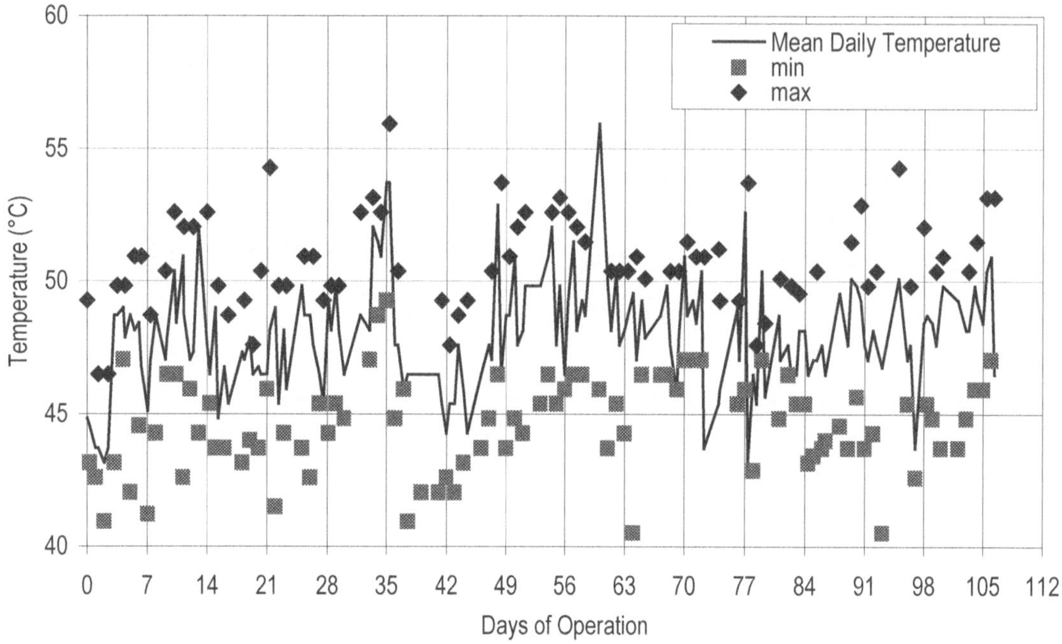

Figure 2-37. Process 3 Temperatures for Thermo 48.

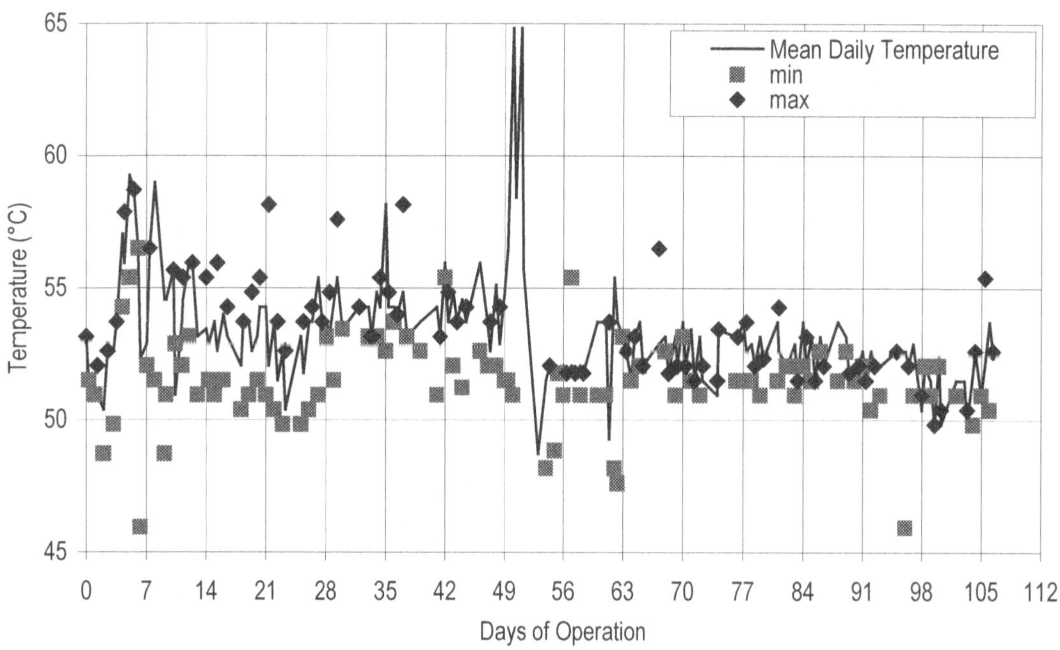

Figure 2-38. Process 3 Temperatures for Thermo 52.

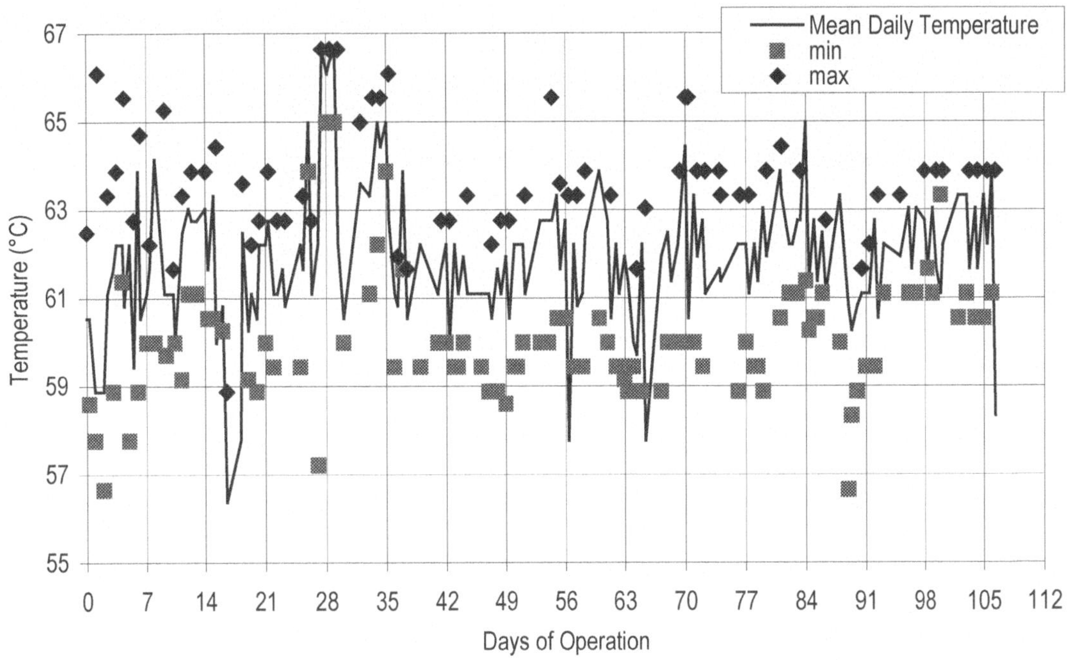

Figure 2-39. Process 3 Temperatures for Thermo 62.

As expected, the Meso Acid digester had the lowest average pH (5.8), the Methane Phase digesters had average pHs in the range from 7.5–8.2 (Figure 2-40), and the lower pH values were measured in the higher temperature thermophilic digesters (Thermo 52 and Thermo 62). Thermo 48 usually showed the highest pH at around 8.3. After sludge feed/withdraw had been stopped for approximately 2 weeks, the Thermo 52 reactor pH increased from approximately 7.8 to 8.3 and remained at this value—a probable indication of increased methane production. On the other hand, the Thermo 62 digester maintained a lower pH for much of the testing period.

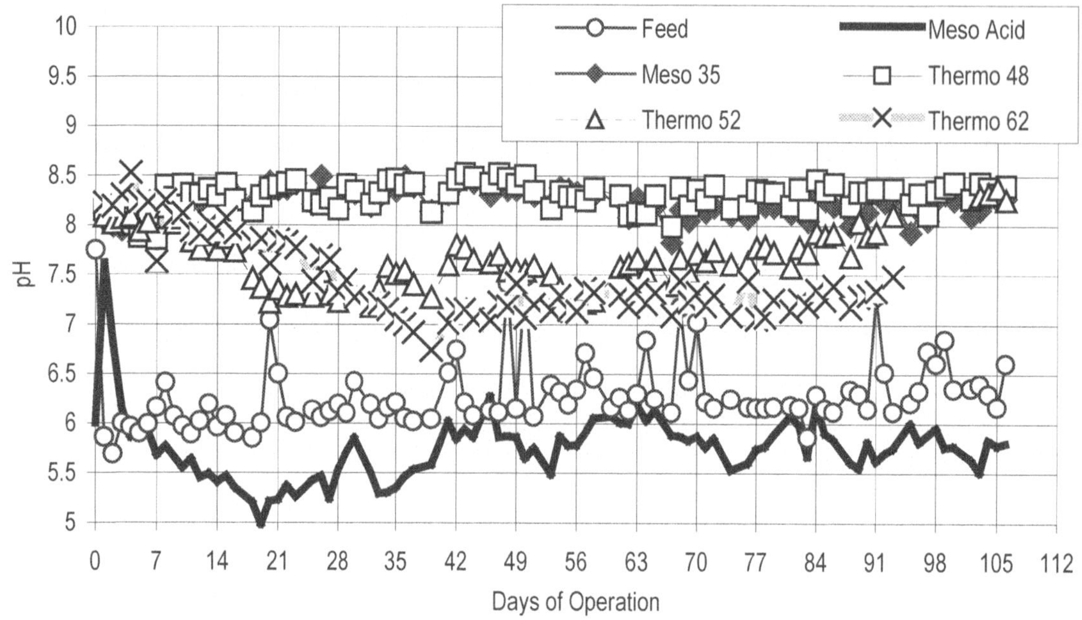

Figure 2-40. Process 3 pH.

As expected from Process 1 results, the TALK of the Meso Acid digester was lowest at around 2,000–3,000 mg/L as $CaCO_3$ (Figure 2-43). The lower temperature methane-phase digesters (Meso 35 and Thermo 48) had the highest TALK levels between 6,000–7,000 mg/L as $CaCO_3$, consistent with the higher pH levels. The Thermo 52 and Thermo 62 digester TALK concentrations increased during the later part of the study but were generally lower (approximately 5,000 mg/L as $CaCO_3$) than the other two methane-phase digesters. During the Process 3 study, digester sludge feed/withdraw was temporarily suspended for the Thermo 52 and Thermo 62 digesters to reduce VFA levels and to increase methane production. During this period, the Thermo 52 reactor TALK increased steadily from approximately 4,000 mg $CaCO_3$/L to approximately 6,000 mg $CaCO_3$/L. The Thermo 62 reactor TALK also increased by more than 1,000 mg $CaCO_3$/L after the sludge feed/withdraw was suspended.

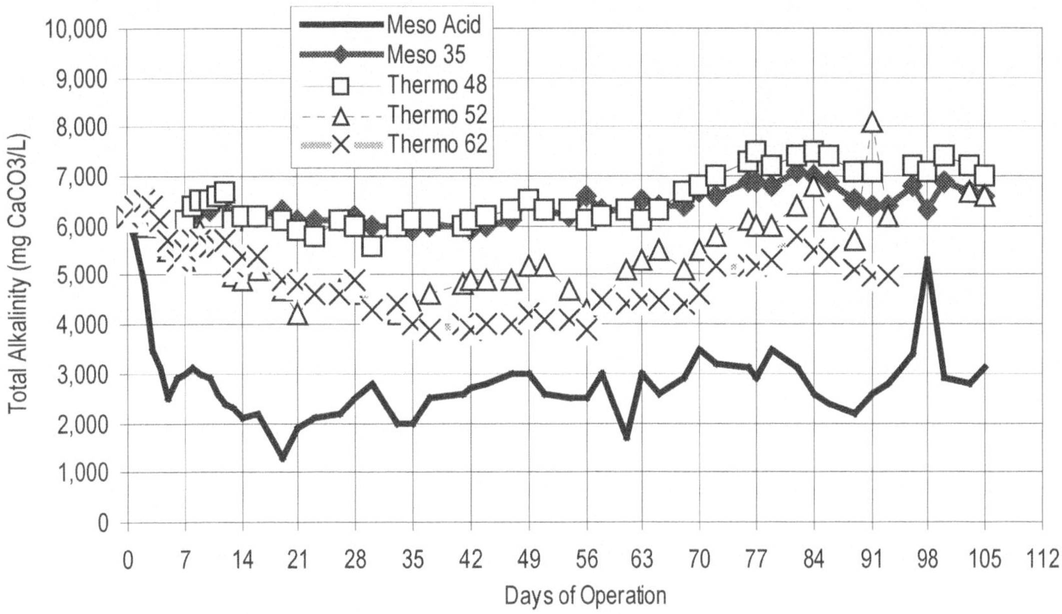

Figure 2-41. Process 3 Total Alkalinity.

Consistent with the pH and TALK results, TVA concentrations in the methane-phase digesters varied with temperature. Following startup, the TVA concentrations in the Meso 35 and Thermo 48 digesters stabilized at relatively low levels (mostly below 1,000 mg/L as acetic acid) (Figure 2-42). Conversely, the Thermo 52 and Thermo 62 digesters showed concentrations well above 5,000 mg/L as acetic acid, similar to the Meso Acid digester. It was difficult to establish viable populations of methanogenic organisms in these two digesters during the Process 3 testing. The TVA/TALK ratios also reflected this difficulty, with TVA/TALK ratios for Thermo 52 and Thermo 62 often being above 1.0, an unstable range for a methane-phase digester (Figure 2-43). These two digesters also had difficulty in degrading propionic acid (Figure 2-44 and Figure 2-45). Figure 2-45 shows similar results to those from the single-stage work (Figure 2-28), in that the specific VFA concentrations increased with increasing thermophilic temperature, and that the thermophilic digester operated at the lowest temperature (48°C) did not have VFA levels significantly different from those in the mesophilic digester (Meso 35).

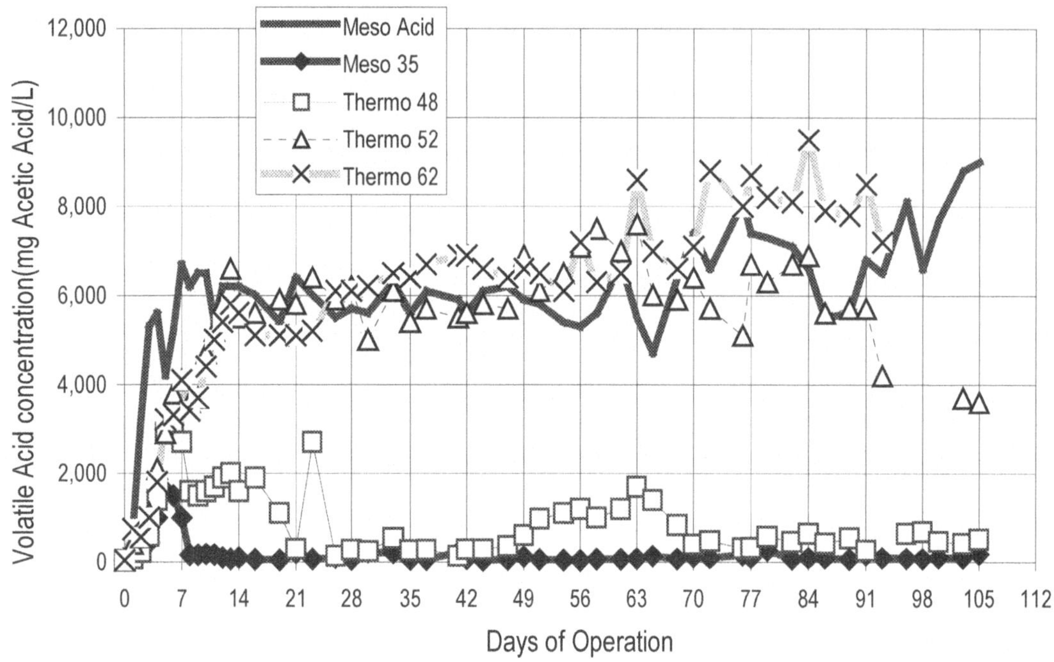

Figure 2-42. Process 3 Total Volatile Fatty Acid Concentration.

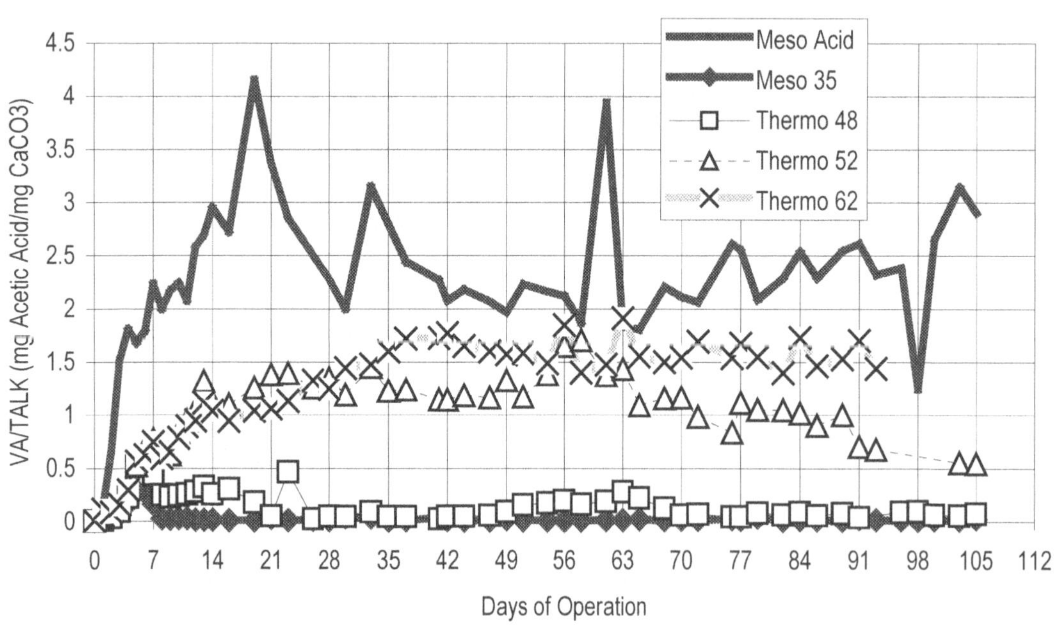

Figure 2-43. Process 3 Total Volatile Acid/Total Alkalinity Ratio (TVA/TALK).

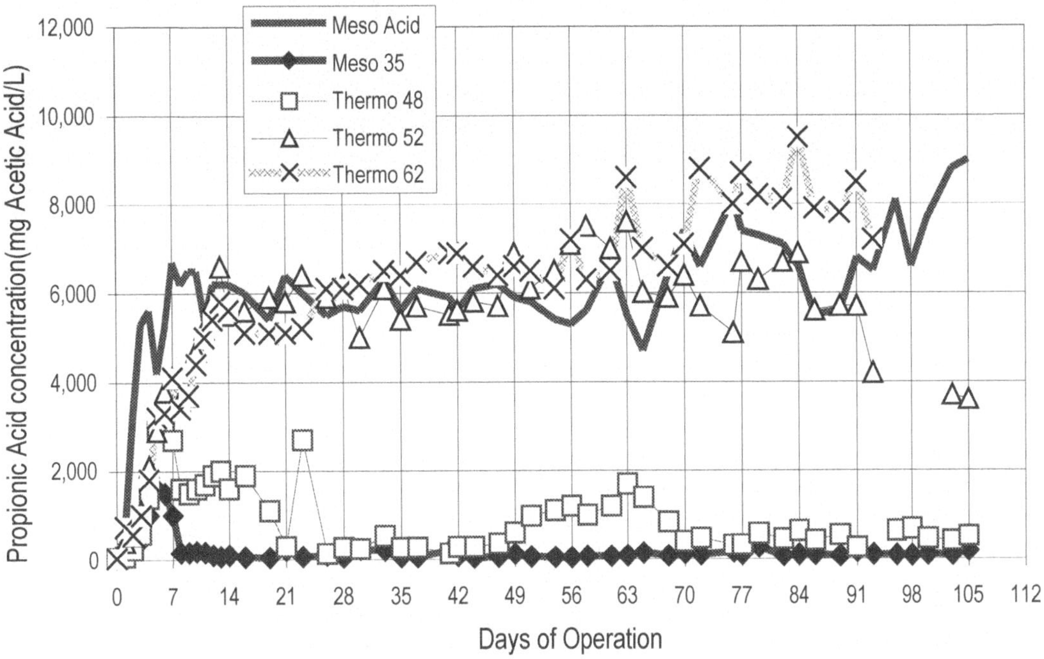

Figure 2-44. Process 3 Propionic Acid Concentration.

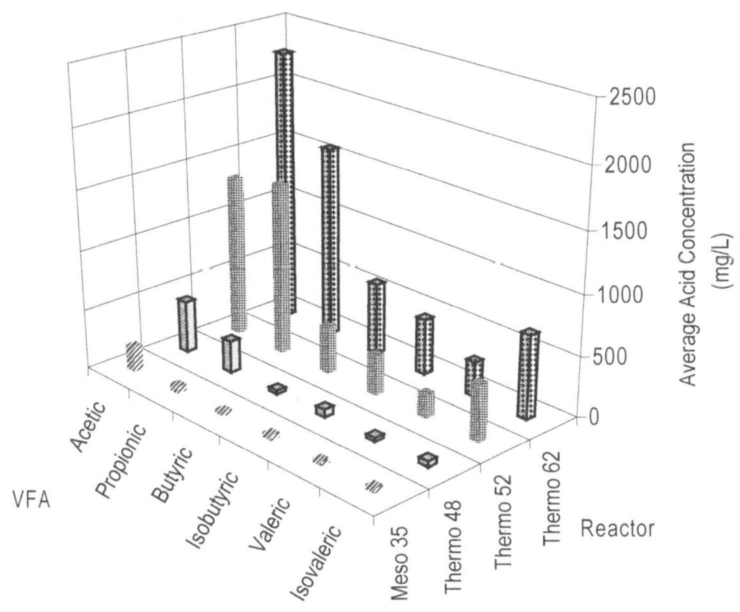

Figure 2-45. Process 3 Averages of Specific Volatile Fatty Acids.

Thermo 48 showed the greatest volatile solids (VS) reduction, followed by Meso 35, Thermo 52, and Thermo 62. The lower VS reductions shown by Thermo 52 and Thermo 62

(Figure 2-46) are consistent with observations regarding the pH, TALK, and acid data, i.e., the performance of these two digesters did not appear as stable as the others. VSR for Meso 35 and Thermo 48 are usually greater than the vector attraction reduction (VAR) requirement of 38%. Of the three thermophilic digesters, Thermo 48 showed the greatest stability, having the lowest TVA/TALK ratio and highest VSR. The higher-temperature digesters did not perform as well.

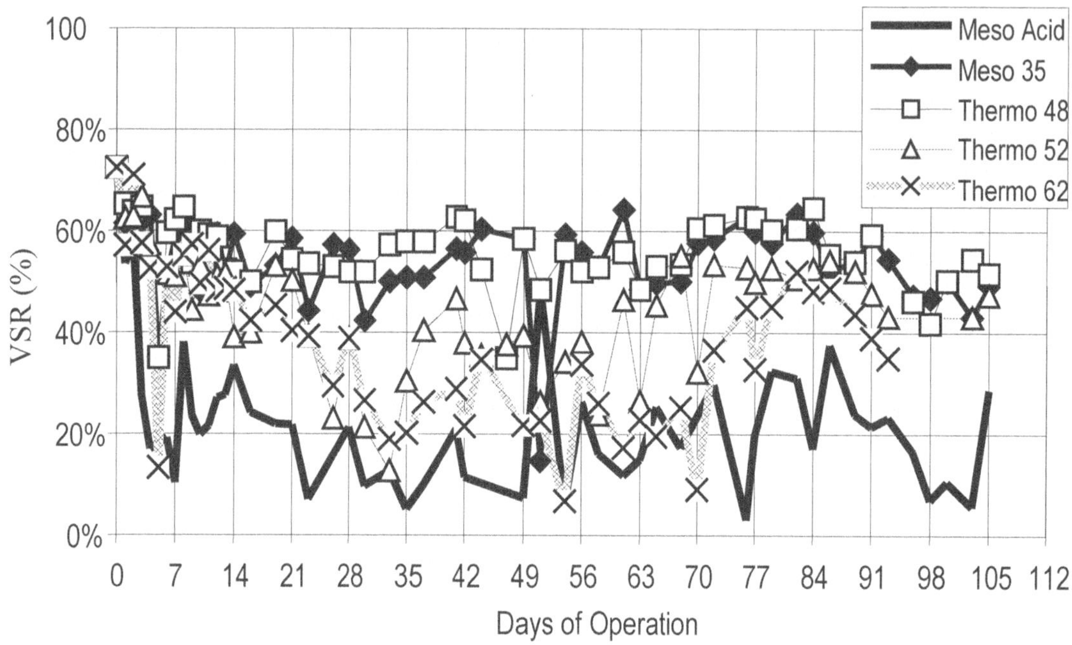

Figure 2-46. Process 3 Volatile Solids Reduction.

Thermo 52 and Thermo 62 consistently reduced fecal coliform concentrations to less than detectable levels (<2 MPN/g TS) (Figure 2-47). Thermo 48 frequently showed less than detectable levels. The arithmetic average fecal coliform concentration in Meso 35 digester effluent was approximately 200 MPN/g TS, less than the Class A fecal coliform criterion of less than 1,000 MPN/g TS. The ability of Meso 35 to achieve this standard was possibly the result of low feed concentrations (average approximately 5.5 log) and/or because it is a two-phase system (i.e., mesophilic acid phase followed by a mesophilic methane phase). The thermophilic digesters achieved between 4-5 logs removal compared to Meso 35, which achieved approximately 3 logs removal (Figure 2-48). The two-stage Meso 35 achieved approximately an extra log removal compared to the single-stage Meso 35 (which achieved approximately a 2-log fecal coliform removal) operated in the last part of the study (see Figure 2-32).

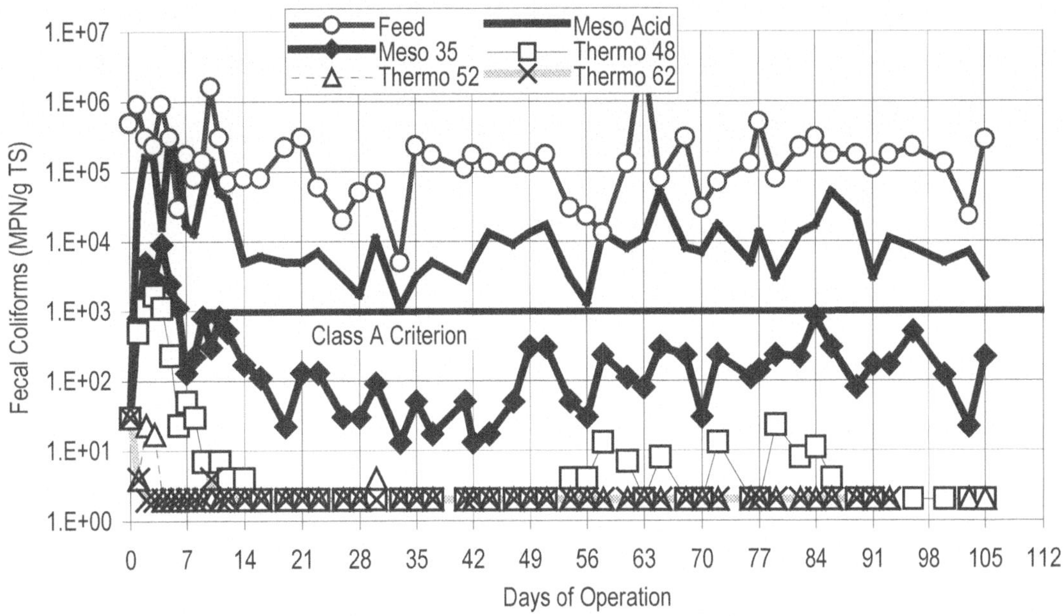

Figure 2-47. Process 3 Fecal Coliforms.

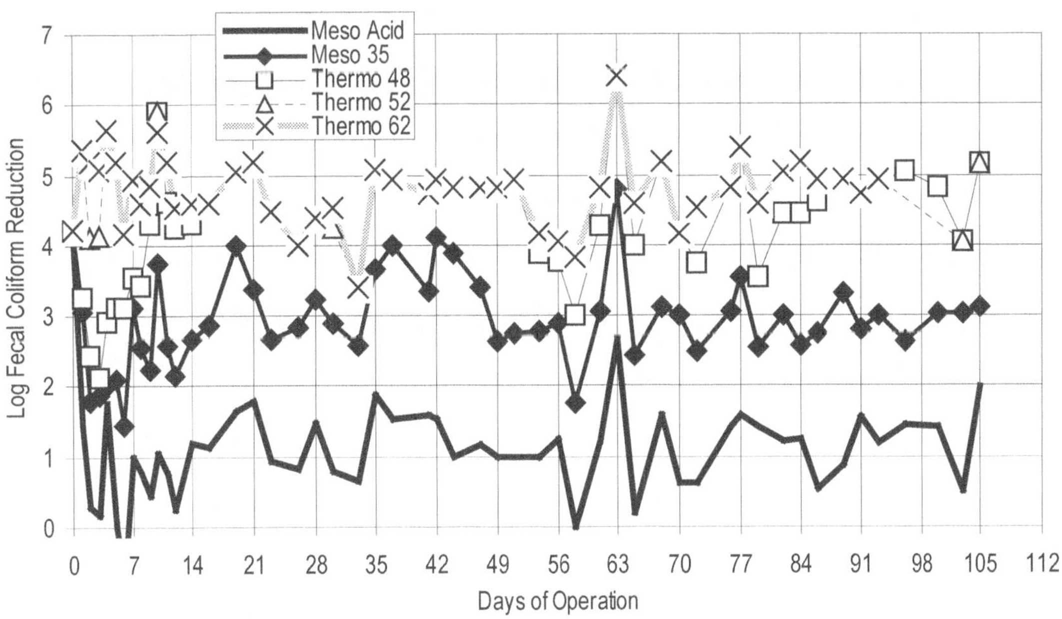

Figure 2-48. Process 3 Fecal Coliform Log Reduction.

In summary, the Thermo 52 and Thermo 62 methane-phase digesters were unable to operate in a stable fashion during this bench-scale study. This was demonstrated by the acidic conditions in the digesters and the low VS destruction of the process. Possible causes for instability include digester temperature variation and the higher operating temperatures. Because

Thermo 48 also experienced temperature variation and was able to operate in a stable fashion, it is likely that the higher operating temperatures (greater than 50° C) were primarily responsible for the instability. Because fecal coliform destruction was consistently excellent throughout the experiment, it appears that thermophilic disinfection is related more to temperature than to digester stability.

2.3.4 Operating Parameters Comparison and Summary

Digester temperature appeared to affect operating performance and stability. For the acid-phase digesters (Process 1), those with the higher operating temperatures (Thermo 54 and 62) were characterized by higher TALK concentrations and pH levels. Although propionic acid concentrations were very high in the acid-phase digesters (>1,200 mg/L), they did not differ substantially at the various operating temperatures.

The same relationship does not appear to hold for the single-stage and methane-phase thermophilic digesters. On the contrary, higher operating temperatures appeared to be associated with higher TVA concentrations and lower TALK. For the single-stage thermophilic digesters (Process 2), TALK and pH were not substantially different between the digesters, but the VFA concentrations, especially propionic acid, were higher at the higher temperatures. The thermophilic methane-phase digesters (Process 3), showed a similar result, although TALK was lower at the higher temperatures.

2.3.5 Fecal Coliform Destruction Comparison and Summary

The primary purpose of the Phase 1 screening evaluation was to determine which process/temperature combinations could achieve the Class A fecal coliform criterion of <1,000 MPN/g TS. Graphical and statistical analyses were performed to evaluate the results of this phase of the study.

There was a consistent inverse relationship between the log mean effluent fecal coliform concentration and digester operating temperature for each process tested (Table 2-4). The log mean effluent fecal coliform concentrations were plotted against mean operating temperatures for each process tested in Phase 1. These means were compared to the Class A criterion of 1,000 MPN/g TS in Figures 2-49, 2-50, and 2-51. The error bars represent the standard deviations.

Table 2-4. Phase 1: Log Mean Effluent Fecal Coliform Concentrations.

Mean Operating Temperature (°C)	Process 1 Acid phase	Process 2 One-Stage	Process 3 Two-stage
35	3.84	2.95	2.42
48	3.34	NA	0.81
49	NA	1.80	NA
52	NA	NA	0.37
53	NA	1.20	NA

Table 2-4. Phase 1: Log Mean Effluent Fecal Coliform Concentrations, continued.

Mean Operating Temperature (°C)	Process 1 Acid phase	Process 2 One-Stage	Process 3 Two-stage
54	2.01	NA	NA
62	1.48	0.72	0.34

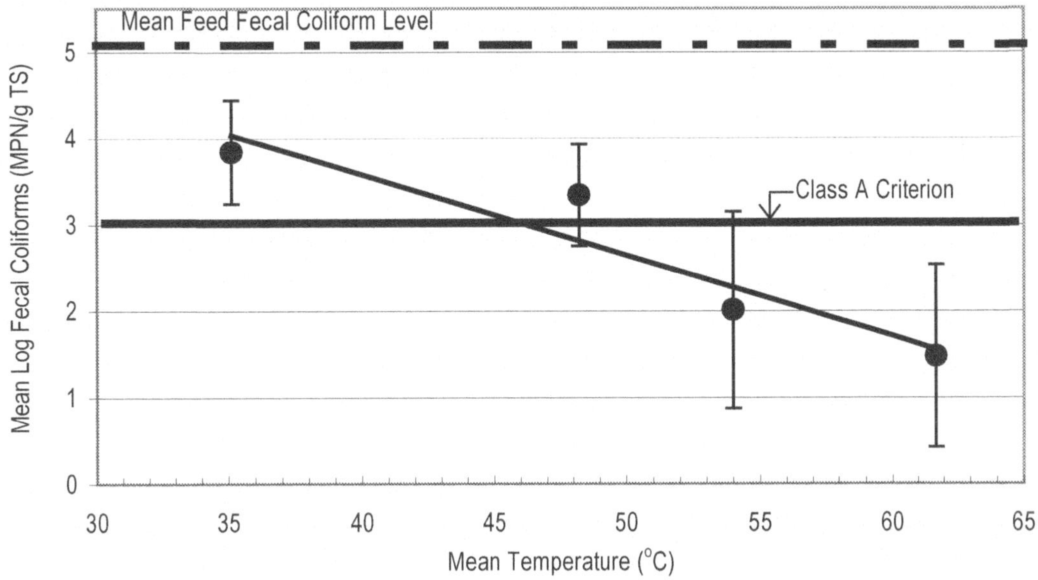

Figure 2-49. Process 1 Thermophilic Acid Phase Mean Effluent Fecal Coliform Concentrations.

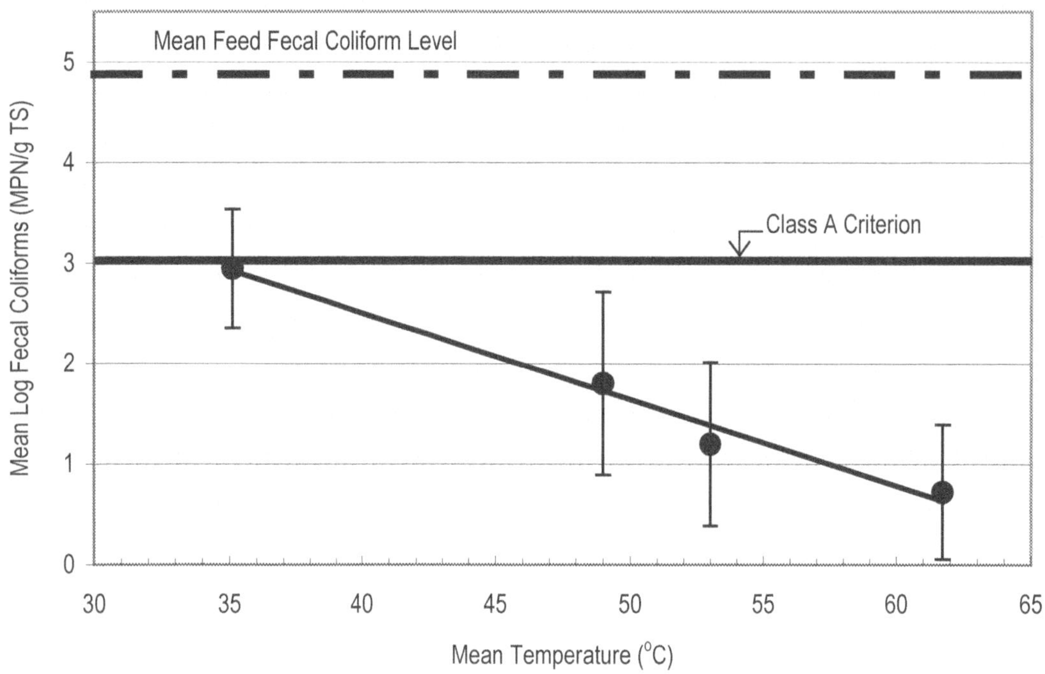

Figure 2-50. Process 2 Single-stage Thermophilic Digesters Mean Effluent Fecal Coliform Concentrations.

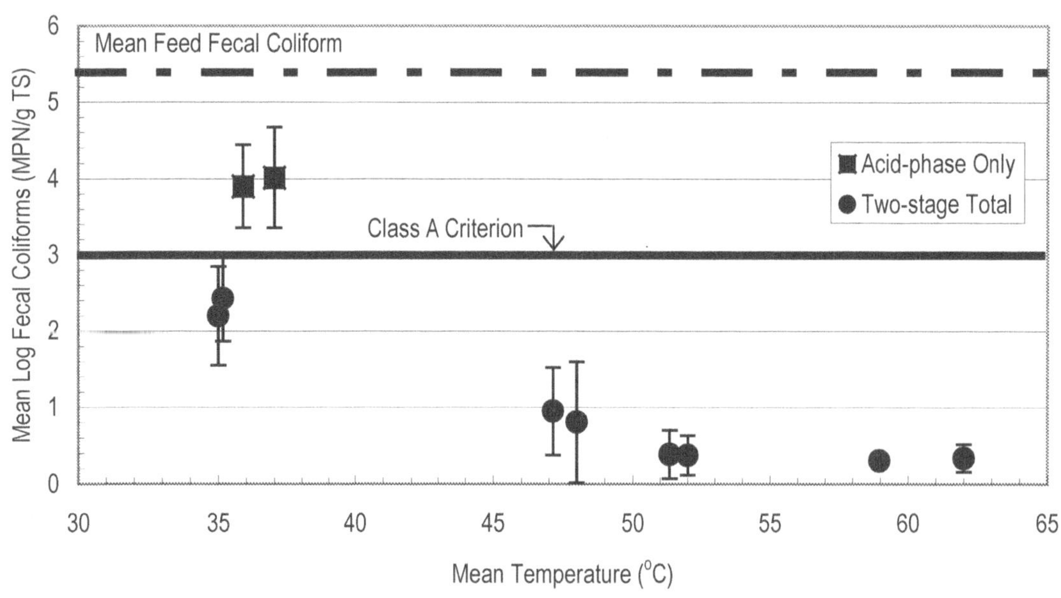

Figure 2-51. Process 3 Two-stage Meso Acid/Thermo Methane Mean Effluent Fecal Coliform Concentrations.

Results for all three processes are plotted in Figure 2-52. The highest fecal coliform reductions were achieved by Process 3, the next highest by Process 2, and the lowest by

Process 1. Because only the thermophilic acid-phase reactor was used to test Process 1 in this phase of the study, it was anticipated that actual Process 1 fecal coliform reductions would be higher than those indicated in Figure 2-52.

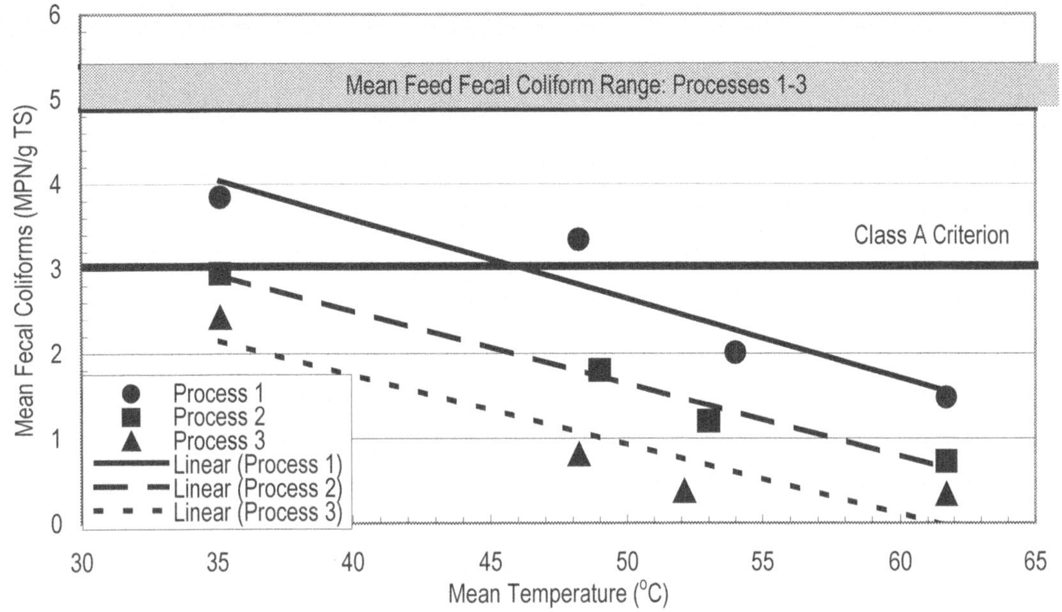

Figure 2-52. Processes 1, 2, and 3 Mean Effluent Fecal Coliform Concentrations.

All three processes proved capable of meeting the Class A fecal coliform requirement at the fecal coliform feed densities occurring during this phase of the project (fecal density means: 160,000–300,000 MPN/g TS, standard deviations: 170,000–860,000 MPN/g TS) as summarized in Table 2-5.

Table 2-5. Phase 1 Results for Meeting Class A Fecal Coliform Requirements.

	Ability to Meet Class A?			
Temperature, °C	35	48	52	62
Process 1 thermo acid phase	No	No	Yes	Yes
Process 2 one-stage	No	Yes	Yes	Yes
Process 3 Meso Acid/Thermo Methane	Yes	Yes	Yes	Yes

In light of these fecal coliform results, all three processes were carried forward to Phase 2 for further evaluation. The thermophilic digesters were operated at a single temperature, 50°C, for comparison. This temperature was chosen because all three processes appeared to meet the Class A fecal coliform requirement at approximately 50°C. The thermophilic anaerobic digestion process, especially for Process 3, appeared to be more stable at <50°C.

2.3.6 Preliminary Sentinel Chamber Test

Preliminary testing was conducted to evaluate the use of sentinel chambers for determining helminth ova destruction. The chambers used for this test were 2-mL plastic cylindrical containers with 30-μm mesh screens at each end. The pore size of the screen allowed liquid to flow through it into the chamber but retained the helminth ova inside the chambers. Preliminary bench-scale tests were conducted to determine the feasibility of using these chambers at full-scale (see also discussion in Section 3.3.1 for Sentinel Chamber tests conducted during Phase 2 of the study).

Chambers containing viable helminth ova were placed sequentially in each phase of the two-phase Process 3 digester systems. Eight chambers were placed in the mesophilic acid stage for 2 days, and then two of the chambers were transferred to each of the four methane-phase digesters for an additional 15 days. The ova in each chamber were examined for viability (Figure 2-53). Approximately 23% of the helminth ova retrieved from the Meso 35 digester were nonviable. This result is not significantly different from the control sample (stored at 4°C for the duration of the test) result of 27% nonviable helminth ova. However, the Sentinel chambers placed in each of the thermophilic methane-phase digesters showed no viable ova after 15 days at thermophilic temperatures. The thermophilic digesters appeared to destroy helminth ova, but the mesophilic digester did not, even after a total of 17 days in a two-phase process. This result suggests that a two-phase mesophilic/mesophilic process will not meet Class A pathogen requirements.

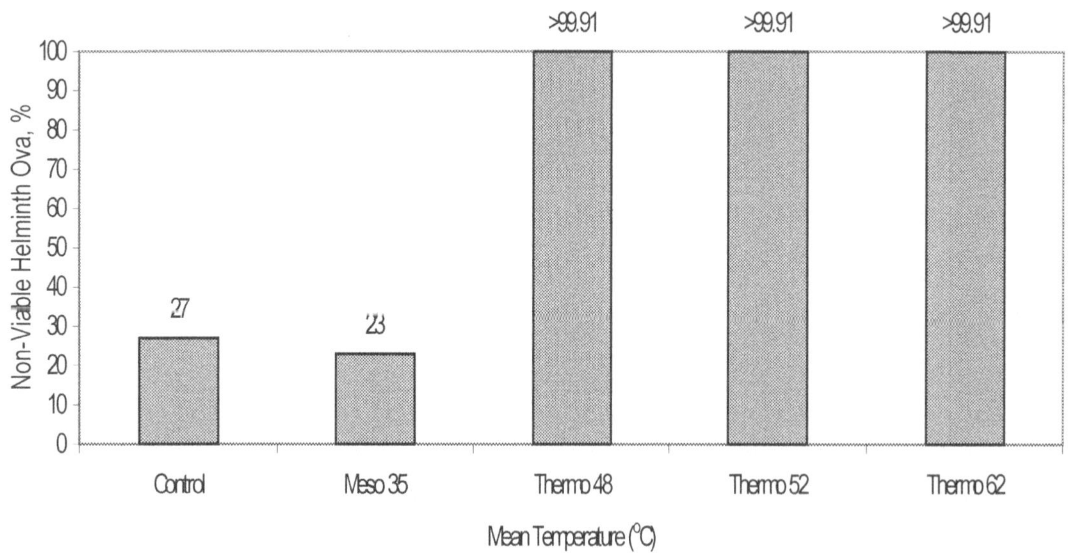

Figure 2-53. Preliminary Helminth Ova Results for Process 3.

CHAPTER 3.0

PHASE 2 PATHOGEN DESTRUCTION

3.1 Purpose

The purpose of Phase 2 was to evaluate promising thermophilic anaerobic digestion processes identified in Phase 1 on the basis of pathogen destruction. Each of the three thermophilic process configurations evaluated in Phase 1 showed they could meet the U.S. EPA Class A fecal coliform criterion and was therefore carried forward to Phase 2. Each process was fed a raw sludge seeded with pathogens to determine their pathogen destruction capabilities. The pathogens seeded were *Salmonella*, poliovirus, and helminth ova. These are the same organisms required for biosolids monitoring in the U.S. EPA 503 regulations for demonstrating production of Class A biosolids. *Escherichia coli* microbes were also seeded to boost the low fecal coliform levels (approximately 10^5) in the EBMUD raw sludge. Thermophilic digesters were operated at 50°C because Phase 1 results indicated that this was the lowest probable temperature where the Class A fecal coliform criterion was consistently met by all three processes.

3.2 Materials and Methods

Four separate digester systems were operated in this part of the study (Figure 3-1):

1. Mesophilic single-stage (Meso 1 or Control) at 35°C;
2. Thermophilic single-stage (Thermo 1) at 50°C;
3. Two-stage, thermophilic acid phase (Thermo AP) at 50°C and mesophilic methane phase (Meso MP) at 35°C (Two-stage Thermo/Meso); and
4. Two-stage, mesophilic acid phase (Meso AP) at 35°C and thermophilic methane phase (Thermo MP) at 50°C (Two-stage Meso/Thermo).

Hydraulic residence times (HRTs) for the acid-phase, methane-phase, and single-stage digesters were 2, 13, and15 days, respectively. Total HRTs were 15 days for each process configuration.

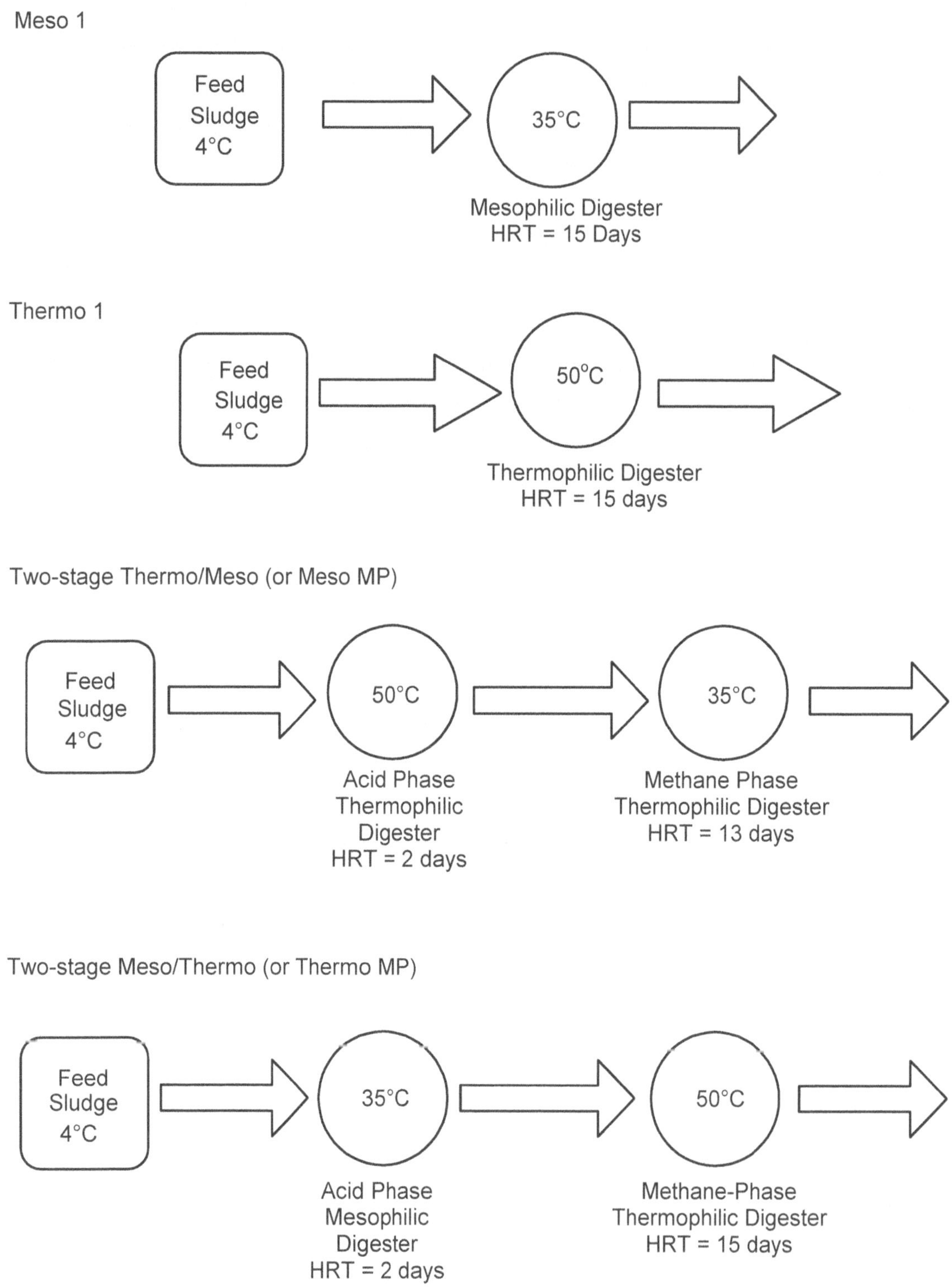

Figure 3-1. Phase 2 Process Configuration.

Erlenmeyer flasks (2-L) with an operating capacity of about 1.5 L were used as digesters (Figure 3-2). Their contents were mixed by magnetic mixers with floating stir bars (VWR, Catalogue No. 58948-482). Preliminary testing performed as part of this study indicated that helminth ova viability was significantly reduced by the grinding action produced by the rotation of a conventional magnetic mixing bar on the bottom of the glass flask (Table 3-1). The use of a floating stir bar (elevated from the flask bottom by a plastic cage) eliminated this problem. The digesters were heated by immersing them in constant temperature water baths equipped with thermostatically-controlled immersion heaters. Digester temperatures were continuously monitored and recorded by Type T thermocouples linked to a PC with data logging software (National Instruments, Austin, TX). Gas production was measured with a wet-tip gas meter (Wet Tip Gas Meter Co., Nashville, TN).

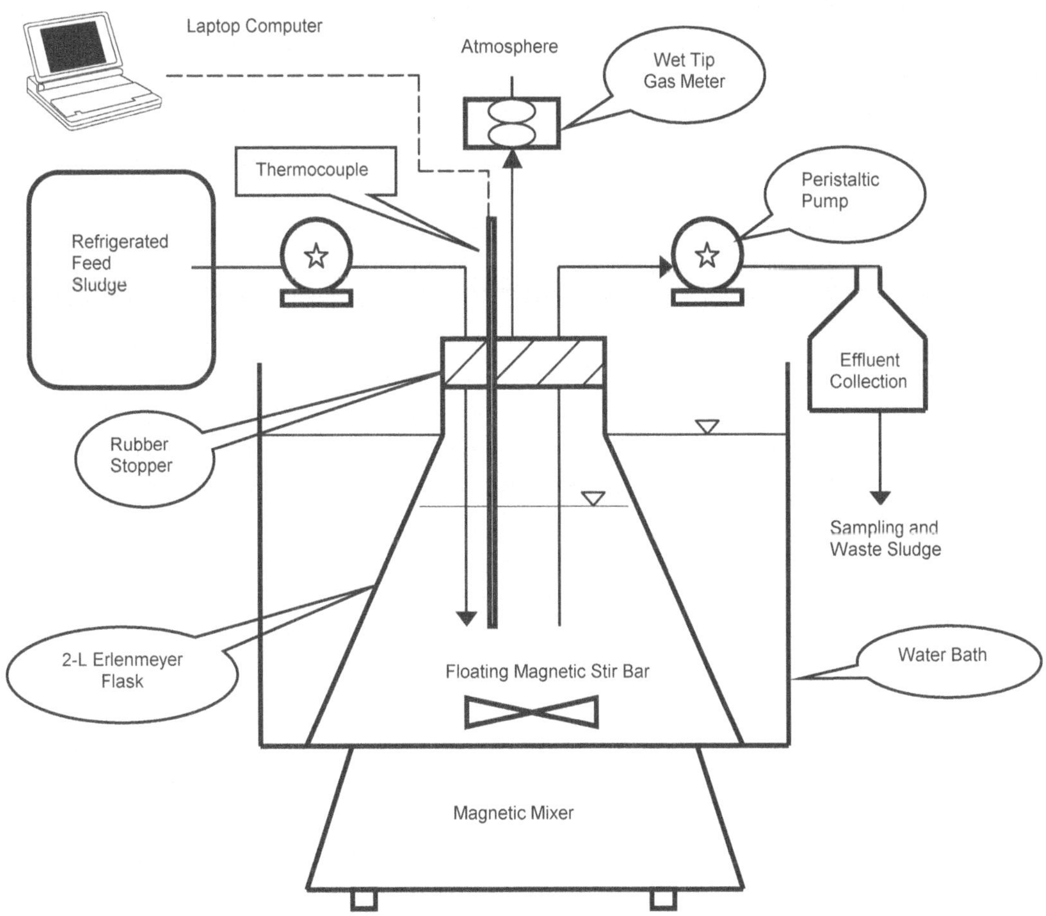

Figure 3-2. Phase 2 Digester Schematic.

Table 3-1. Stir Mechanism Effect on Helminth Ova Concentrations (No. viable helminth ova seen per 10 ml).

Stir Mechanism	Time = 0 hours	Time = 24 hours
Conventional Stir Bar	44	0
Floating or caged stir bar	46	51
No stir bar	49	50

Peristaltic pumps were used for feeding and withdrawing sludge from the reactors. HRTs were controlled by adjusting flow rates and pumping intervals. Feeding was semi-continuous on a draw/fill/hold cycle. The draw and fill portions of the cycle were each approximately 1–2 minutes—relatively short compared to the hold cycle. The hold portion of the cycle ranged from 1–6 hours depending upon the digester HRT. Digester effluent was stored in a container cooled to approximately 10°C and capped with a gas seal that allowed displaced gas to escape but prevented air from entering.

Digester feed was primary sludge (PS) from the EBMUD main wastewater treatment plant. Prior to use, the PS was filtered through a 5 × 5 mesh per lineal inch (63.2% open area) wire screen and then through two sieves (No. 16 with 1.18-mm opening and No. 70 with 0.212-mm opening). The filtered feed sludge was allowed to settle overnight, decanted, and stored in a refrigerator at approximately 10°C and was mixed with a paddle stirrer prior to each feeding cycle. Filtering and decanting was used to prevent pump and tube plugging. Thickened waste activated sludge (TWAS) was not used for feed in Phase 2 because of nocardioform foaming in the digesters.

The feed sludge was seeded with fecal coliforms (*E. coli*) and three pathogens: helminth ova, *Salmonella*, and poliovirus. Helminth ova (*Ascaris suum*) (Excelsior-Sentinel, Ithaca, NY) were collected from pig feces. A wild strain of *Salmonella* was extracted from EBMUD raw sewage and cultured by BioVir Laboratories, Inc., Benecia, CA. The wild-type strain had a higher survival rate than standard ATCC strains. Poliovirus Type 2-vaccine strain (Source: California Dept. of Health Services) was used for the enteric virus. At the beginning of the study, both the feed sludge and digesters were spiked to pathogen concentrations that met or exceeded the U.S. EPA requirements (U.S. EPA, 1989) to ensure that pathogen concentrations reached equilibrium immediately and that pathogen destruction could be evaluated from the very start of the test. The alternative would have been to seed only the feed sludge and wait 3–4 HRTs before reaching equilibrium. The study duration was 3 HRTs (45 days).

In addition to seeding the feed sludge directly, helminth ova destruction was evaluated using Sentinel Chambers (Excelsior-Sentinel, Ithaca, NY). Helminth ova concentrations inside the vials were compared to those in the digesting sludge.

Analytical methods from *Standard Methods for the Examination of Water and Wastewater* (APHA, 1998) were used for the sludge analysis. Pathogen testing was by Pathogen Equivalency Committee (PEC) approved methods (U.S. EPA, 1999).

3.3 Results and Discussion

Results are reported for the following digesters:

1. Mesophilic single-stage (Control or Meso 1);
2. Thermophilic single-stage (Thermo 1);
3. Mesophilic acid-phase (Meso AP);
4. Thermophilic acid-phase (Thermo AP);
5. Mesophilic methane phase (Meso MP); and
6. Thermophilic methane phase (Thermo MP).

The Meso MP and Thermo MP results represent the effect of an entire process (including the acid phase digesters). Digester effluent pathogen samples (24-hour composite) were collected three times per week. Feed sludge pathogens and effluent fecal coliform concentrations were measured weekly.

Because the digester cultures were seeded with pathogens at the beginning of the study, the pathogen concentrations in the first samples drawn from the digesters were very close to those in the feed sludge. A rapid decline in effluent pathogen concentrations was observed in most of the digester effluents during the first few days of the test.

3.3.1 Fecal Coliform and Pathogen Destruction

Average feed sludge fecal coliform concentrations were approximately 10^5 MPN/g TS during Phase 1 of this study (see Chapter 2.0). These concentrations may be lower than those in raw sludges at other municipal wastewater treatment plants, probably because of the use of prechlorination at the EBMUD plant headworks for odor control. To compensate for the relatively low background fecal coliform concentrations, the feed sludge for Phase 2 was spiked with *E. coli* to yield fecal coliform levels of between 10^6 and 10^7 MPN/g TS in the sludge. Higher feed sludge concentrations would provide the appropriate challenge to the processes for meeting the Class A criterion of less than 1,000 MPN/g TS.

Digester effluents from Thermo 1, Meso MP, and Thermo MP had fecal coliform concentrations consistently lower than the Class A fecal coliform criterion of 1,000 MPN/g TS (Table 3-2, Figure 3-3). The Thermo MP digester effluent had the lowest fecal coliform concentrations. The control digester (Meso 1) only had one value less than the Class A criterion. These results are consistent with those from Phase 1 although the feed sludge concentrations were 1–2 orders of magnitude greater.

Table 3-2. Phase 2 Fecal Coliform Results (MPN/g TS).

Days of Operation	Meso AP	Thermo AP	Meso MP	Thermo MP	Meso 1	Thermo 1	Feed
1	220	$1.3x10^3$	110	<2	$9x10^4$	17	$17x10^6$
8	$2.2x10^4$	17	33	4	$2.4x10^4$	4	$2.2x10^6$
15	$9x10^4$	110	130	4	$1.6x10^5$	34	$9x10^6$
22	90	<2	17	<2	230	7	$3x10^6$
29	500	2	500	<2	$2.4x10^4$	17	$3x10^6$
36	500	$1.6x10^4$	26	2	$3x10^4$	NA	$1.6x10^7$
43	$1.6x10^5$	$3x10^3$	4	<2	$3x10^4$	23	$3.5x10^6$

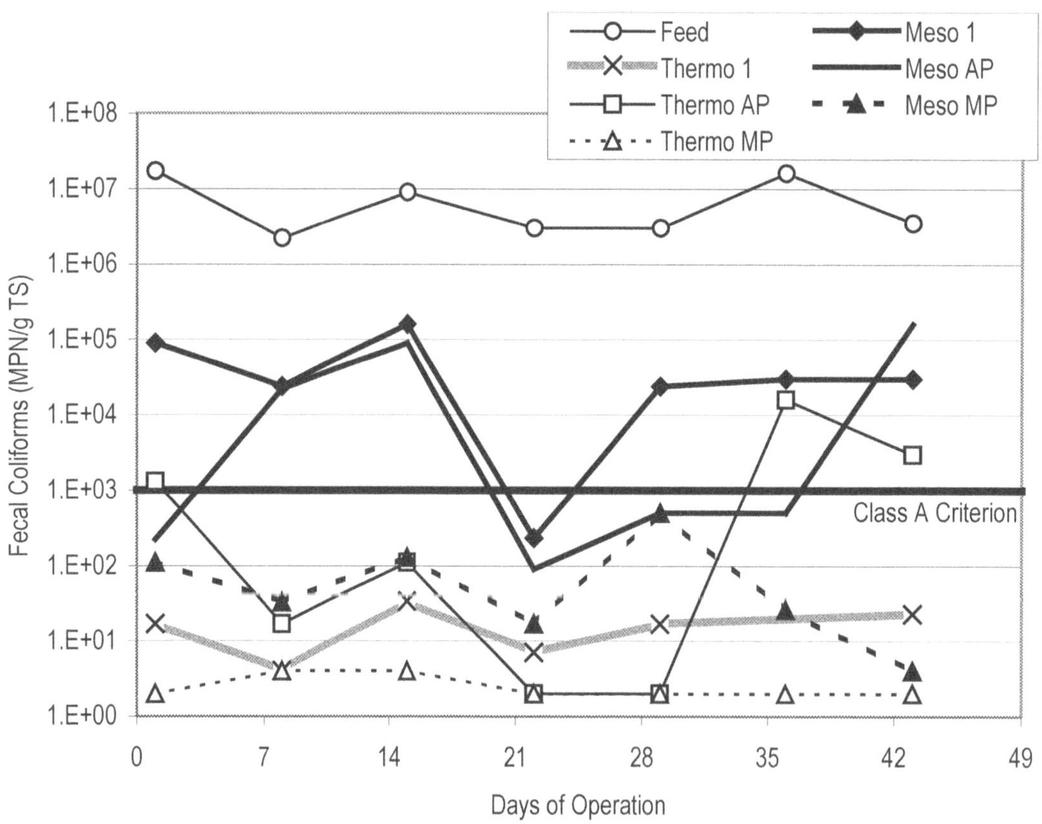

Figure 3-3. Fecal Coliform Concentrations.

Alternatives 3 and 4 of the 40 CFR Part 503 regulations allow monitoring of fecal coliforms or *Salmonella* to judge compliance with the Class A sludge requirements. Phase 1 results suggested that one or more of the thermophilic processes would achieve the Class A fecal coliform requirement. The feed sludge was spiked with *Salmonella* to concentrations of approximately 10^8 MPN/4 g TS—well in excess of the U.S. EPA minimum spike requirement of at least 10^3 MPN/g TSS (U.S. EPA, 1989). Previous studies have reported raw sludge *Salmonella* concentrations that were much closer to the U.S. EPA minimum spike requirement than to those observed here (Garber, 1977, 1982; Stern and Farrell, 1977). Although it reduced the chances of meeting the Class A requirements, spiking at the higher concentration allowed for identification of a *Salmonella* destruction endpoint and a comparison of the relative effectiveness for *Salmonella* destruction.

None of the digester effluents consistently achieved the Class A *Salmonella* criterion of <3 MPN/4 g TS at these extremely high seed levels (Table 3-3, Figure 3-4). However, the three thermophilic processes (Thermo 1, Meso MP, and Thermo MP) consistently exhibited log reductions of between 3 and 7 log units. The Thermo MP digester showed the greatest reductions and met the Class A *Salmonella* criterion most often.

Table 3-3. Phase 2 *Salmonella* Results (MPN/ 4 g TS).

Days of Operation	Meso AP	Thermo AP	Meso MP	Thermo MP	Meso 1	Thermo 1	Feed
0	2.4×10^{8}	2.4×10^{7}	1.1×10^{8}	4.6×10^{7}	1.1×10^{8}	1.1×10^{8}	4.6×10^{7}
4	$>1.6\times10^{3}$	$>1.6\times10^{3}$	240	4	$>1.6\times10^{3}$	460	
7	$>1.6\times10^{4}$	150	$>1.6\times10^{4}$	2	$>1.6\times10^{4}$	$>1.6\times10^{4}$	9.3×10^{7}
9	1.1×10^{6}	120	3.6×10^{3}	460	1.6×10^{6}	2	
11	4.3×10^{4}	23	<2	23	9.3×10^{4}	2.1×10^{4}	
14	2.4×10^{5}	240	$>1.6\times10^{3}$	23	$>1.6\times10^{4}$	1.5×10^{4}	2.4×10^{8}
16	1.1×10^{6}	$>1.1\times10^{3}$	1.1×10^{4}	93	4.6×10^{5}	4.6×10^{5}	
18	1.1×10^{6}	75	1.1×10^{3}	9	4.3×10^{4}	7.2×10^{3}	
21	9.1×10^{3}	23	$>1.6\times10^{3}$	<3	9.1×10^{3}	<3	2.3×10^{8}
23	290	23	30	<3	610	3	
25	<3	15	<3	<3	360	<3	
28	1.1×10^{3}	23	$>1.1\times10^{3}$	9	$>1.1\times10^{3}$	43	$>1.1\times10^{3}$
30	280	9	150	$>1.1\times10^{3}$	91	nd	
32	2.1×10^{3}	23	2.1×10^{3}	<3	$>1.1\times10^{3}$	240	
35	$>1\times10^{4}$	$>1.1\times10^{3}$	$>1\times10^{4}$	1.1×10^{3}	1×10^{6}	1.1×10^{3}	4.3×10^{7}
39	$>1\times10^{4}$	23	<30	<3	2.1×10^{5}	93	
44	$>1.1\times10^{4}$	1.1×10^{3}	23	<3	9.3×10^{3}	43	

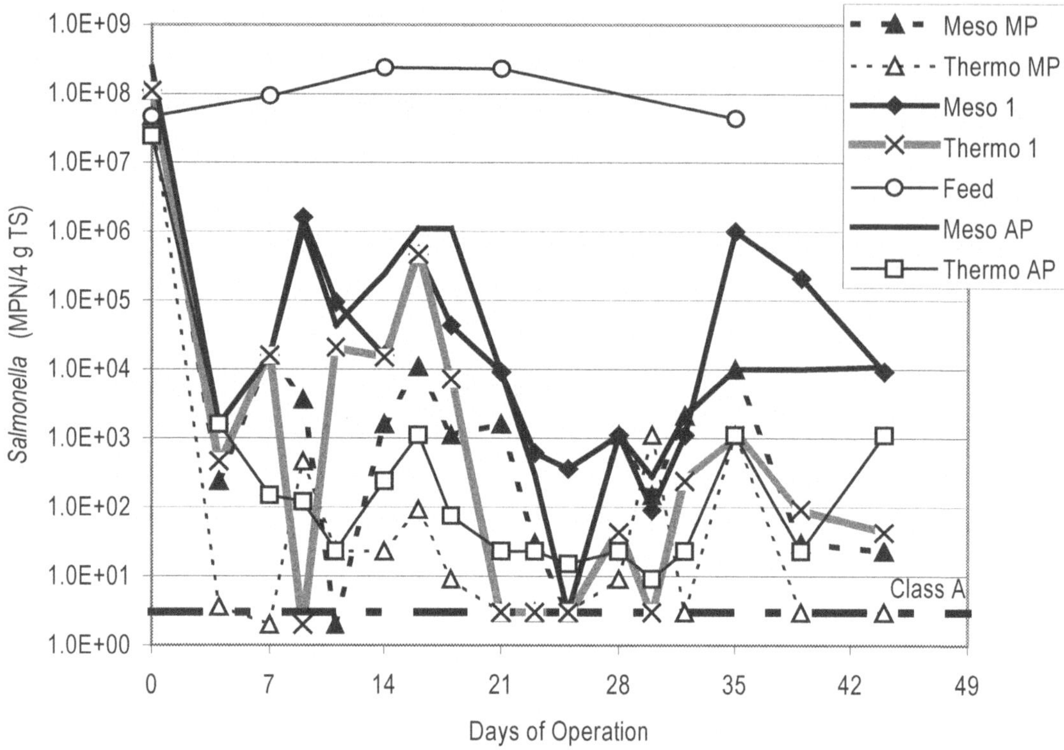

Figure 3-4. ***Salmonella*** **Concentrations.**

The feed sludge was spiked with poliovirus to meet the U.S. EPA minimum spike requirement of 10^3 PFU/g TSS for enteric viruses (U.S. EPA, 1989). Feed sludge enteric virus concentrations were approximately 10^3–10^4 PFU/4 g TS (Table 3-4, Figure 3-5). The control digester produced approximately a 1-log reduction. The Meso MP digester showed between 2 and 4 logs reduction but did not consistently meet the Class A requirement for enteric viruses (<1 plaque forming units (PFU)/4 g TS). Both the Thermo 1 and Thermo MP digesters achieved the Class A requirement with only one or two exceptions—Thermo MP on Day 35 and Thermo 1 on Days 7 and 11. Poliovirus results in Table 3-4 show spiking/recovery data on 2 days (Days 16 and 30). These spikes were performed to confirm recovery rates of poliovirus in the various digester sludges. Recovery rates were determined by spiking a matrix (in this case digester sludge) with a known quantity of a specific organism and comparing the measured concentration with the predicted or calculated value. Recovery rates for poliovirus ranged from <1% to 85% but are usually approximately 20%, which is considered to be a typical recovery rate from this matrix (personal communication with BioVir Laboratories, Inc.).

Table 3-4. Phase 2 Poliovirus Results (PFU/ 4g TS).

Days of Operation	Meso AP	Thermo AP	Meso MP	Thermo MP	Meso 1	Thermo 1	Feed	Comments
0	10,000	5,000	3,000	1,300	3,000	90	1,000	initial spike: 6 x 10E4/ mL
4	300	30	25	0	170	0		
7	600	68	14	0	211	3	700	
9	388	66	5	0	170	0		
11	440	108	2	0	62	1		
14	124	46	1	0	47	0	1,200	
16	2,100	30	2	0	88	0		
16	*37,000*	*20,000*	*146*	*44,000*	*41,000*	*45,000*		*spike: 2 x 10E5/ mL*
18	154	28	4	0	87	0		
21	180	12	0	0	50	0	840	
23	230	34	3	0	35	0		
25	320	88	0	0	60	0		
28	300	170	0	0	9	0	7,000	
30	290	130	0	nd	100	0		
30	*90,000*	*10,000*	*20,000*	*nd*	*170,000*	*170,000*		*spike: 2 x 10E5/ mL*
36	360	86	7	0	75	0		
35	200	140	7.5	2	70	0	900	
39	170	41	11	0	80	0		
44	270	66	3	0	50	0		

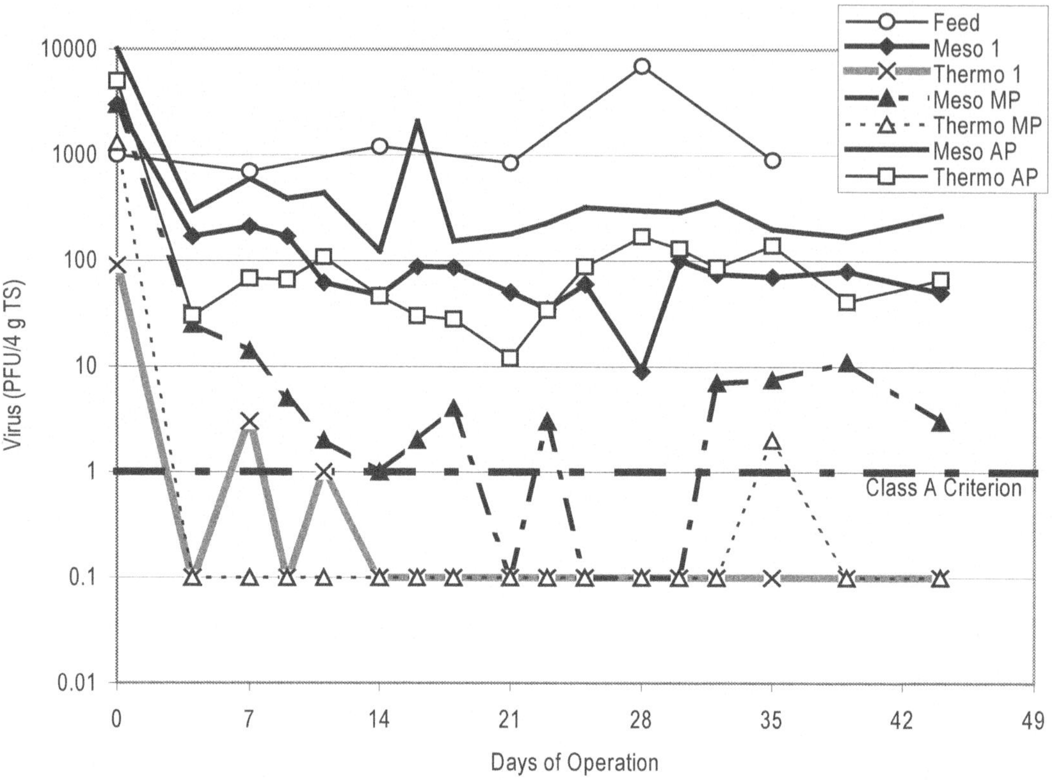

Figure 3-5. Poliovirus Concentrations.

Feed sludge was spiked with helminth ova to meet the U.S. EPA minimum spike requirement of 10^2 ova/g TSS for total helminths (U.S. EPA, 1989). Results here are reported as "percent viability," and 0% viability is considered equivalent to the Class A criterion of <1 ova/4 g TS (Table 3-5, Figure 3-6). Feed sludge viabilities are approximately 90% with one exception. The control digester viabilities fluctuated widely (between 25% and 96%) but never achieved the Class A criterion. The three thermophilic processes consistently achieved levels of less than 5% viability. The Thermo MP appeared to yield the best performance—the viability fell to zero within 4 days, and there was only one value in excess of the Class A criterion. Similarly, viabilities in the Thermo 1 dropped to zero within 4 days but there were three values in excess of the Class A criterion. Meso MP values decreased more slowly after the initial seeding and took about 2 weeks before reaching an apparent equilibrium of 5% viability. The Meso MP digester exhibited consistently low levels (<5%) however, it only achieved the Class A criterion twice over the study period. It appeared that most of the helminth ova destruction in this two-stage process occurred in the first stage (thermophilic acid phase or AP). In fact, the Thermo AP digester achieved the Class A criterion 10 times over the study period and often had lower viability values than the Meso MP digester. It is likely that most of the helminth ova destruction occurs in the Thermo AP digester, and that the viable ova found in the second stage Meso MP effluent represent washout from the initial digester seeding.

Table 3-5. Phase 2 Helminth Results (% Viable).

Days of Operation	Meso AP	Thermo AP	Meso MP	Thermo MP	Meso 1	Thermo 1	Feed
0	95	80	88	95	95	100	100
4	89	0	87	0	86	0	
7	83	0	52	0	90	0	88
9	76	1	44	0	96	0	
11	80	0	5	0	80	13	
14		5	5	0	40	1	35
16	35	0	6	0	55	0	
18	40	25	5	0	85	0	
21	20	0	5	0	25	0	90
23	55	0	2	0	52	0	
25	2	0	3	0	80	0	
28	52	0	1	0	96	0	85
30	75	1	1		55	0	
32	45	0	0	0	48	0	
35	60	30	4	17	80	3	90
39	80	0	1	0	64	0	
44	90	5	0	0	50	0	

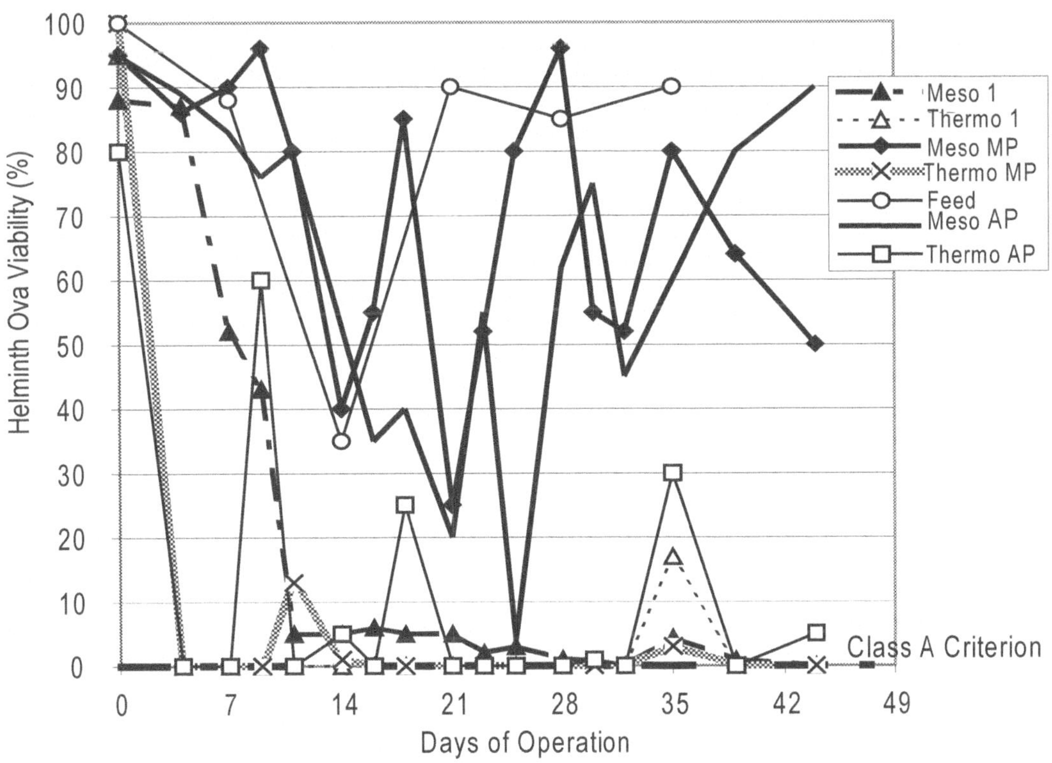

Figure 3-6. Helminth Ova Viability.

Concurrent with feed sludge seeding, five or six Sentinel Chambers were placed in the Thermo 1 and Thermo AP/Meso MP digesters. Individual chambers were retrieved at various times from 6 hours to 10 days to determine the rate of helminth destruction. For the two-stage process, chambers were placed in the acid-phase digester first and then moved to the methane phase digester to simulate a two-phase process. For example, in the two-stage Thermo AP to Meso MP process (Trial 1), six chambers were placed in the Thermo AP digester for 0.25 days. A period of 0.25 days was selected to determine helminth reduction at a time period less than the average 2-day HRT, since not all of the digester sludge remains in a continuous-flow stirred tank reactor (CSTR) for a full HRT. All six were withdrawn and one was analyzed. The other five were inserted into the Meso MP digester and withdrawn at various times (at 3, 5, 7, and 10 days) for analysis. With the exception of the chamber withdrawn after 0.25 days from the Thermo AP digester in Trial 1, no viable helminth ova were detected in any of the chambers (Table 3-6). For the chamber that spent 6 hours in the Thermo AP digester, the helminth ova viability was 5%. This observation indicates that substantial ova kill occurred in a thermophilic acid digester at 50°C within the first 6 hours. Two control chambers stored under refrigeration during the entire test period showed viabilities of 92% and 97%.

Table 3-6. Sentinel Chamber Results.

	Control (held in refrigerator)				
removed	Day 52	Day 52			
% viable	92	97			
	Thermo AP to Meso MP - Trial 1				
inserted Thermo AP	Day 10				
removed Thermo AP	Day 10				
elapsed time (d)	0.25				
% viable	5				
transferred to Meso MP	Day 10	Day 10	Day 10	Day 10	Day 10
removed Meso MP	Day 14	Day 16	Day 18	Day 18	Day 21
elapsed time in Thermo AP (d)	0.25	0.25	0.25	0.25	0.25
elapsed time in Meso MP (d)	3.83	5.83	7.80	7.80	10.78
% viable	0	0	0	0	0
	Thermo AP to Meso MP - Trial 2				
inserted Thermo AP	Day 24				
removed Thermo AP	Day 25				
elapsed time (d)	0.75				
% viable	0				
transferred to Meso MP	Day 25	Day 25	Day 25	Day 25	Day 25
removed Meso MP	Day 28	Day 30	Day 32	Day 32	Day 35
elapsed time in Thermo AP (d)	0.75	0.75	0.75	0.75	0.75
elapsed time in Meso MP (d)	3.00	5.06	7.07	7.07	10.03
% viable	0	0	0	0	0
	Single-stage Thermo 1- Trial 1				
inserted Thermo 1	Day 14	Day 14	Day 14	Day 14	Day 14
removed Thermo 1	Day 16	Day 18	Day 21	Day 21	Day 23
elapsed time (d)	1.97	3.96	6.94	6.94	8.98
% viable	0	0	0	0	0

Table 3-6. Sentinel Chamber Results, continued.

	Single-stage Thermo 1 - Trial 2				
inserted Thermo 1	Day 25	Day 25	Day 25	Day 25	Day 25
removed Thermo1	Day 28	Day 30	Day 32	Day 32	Day 35
elapsed time (d)	3.02	5.03	7.05	7.05	10.01
% viable	0	0	0	0	0
	Thermophilic Water Bath				
inserted Thermo bath	Day 16	Day 16	Day 16	Day 16	
removed Thermo bath	Day 18	Day 21	Day 23	Day 25	
elapsed time (d)	1.97	4.98	7.02	8.95	
% viable	0	0	0	0	

3.3.2 Digestion Process Performance

Standard operating parameters for assessing digester process stability and performance were collected weekly, except for temperature (continuous), pH and gas production (daily). Weekly samples were 24-hour composites (e.g., VFA, TALK, VS, and TS).

There are spikes in the HRTs for single-stage and methane-phase digesters (Figure 3-7). The spikes reflect brief periods where feed or discharge rates were temporarily altered to balance digester volume, or when feeding was temporarily suspended to stabilize digester chemistry. Generally, HRTs are close to the target values of 2, 13, and 15 days for the acid-phase, methane-phase, and single-stage digesters, respectively.

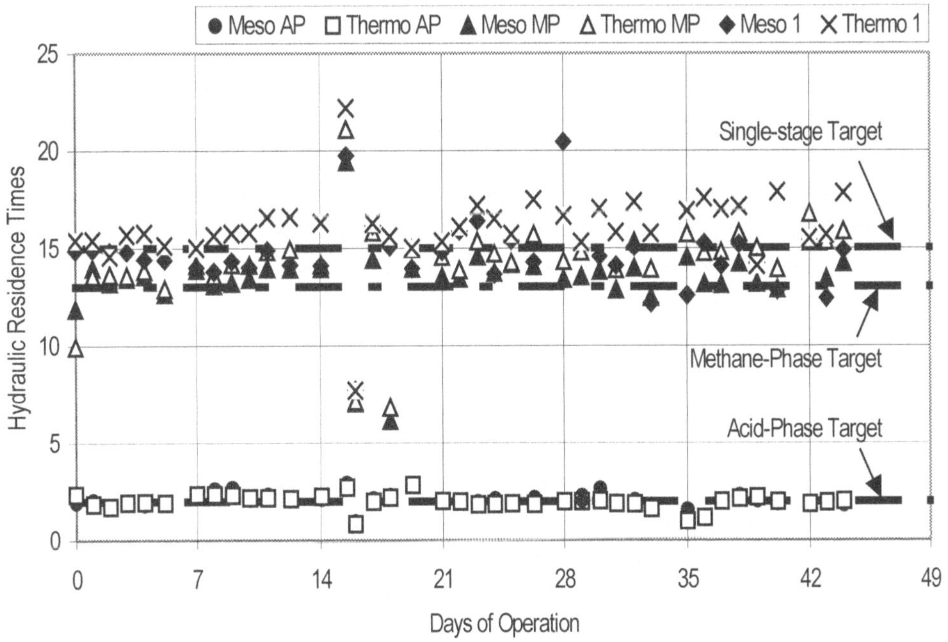

Figure 3-7. Hydraulic Residence Times.

Temperatures were measured and recorded continuously for all digesters. Average temperatures were within 0.3°C of the target values of 35°C and 50°C, for mesophilic and thermophilic digesters, respectively. Digester pH values were measured daily and average values were within the ranges expected: between 5.5 and 6.2 for acid-phase digesters and 7.5 and 8.0 for methane-phase and single-stage digesters (Figure 3-8). However, Meso 1 and Meso MP had pHs < 7.0 during several periods.

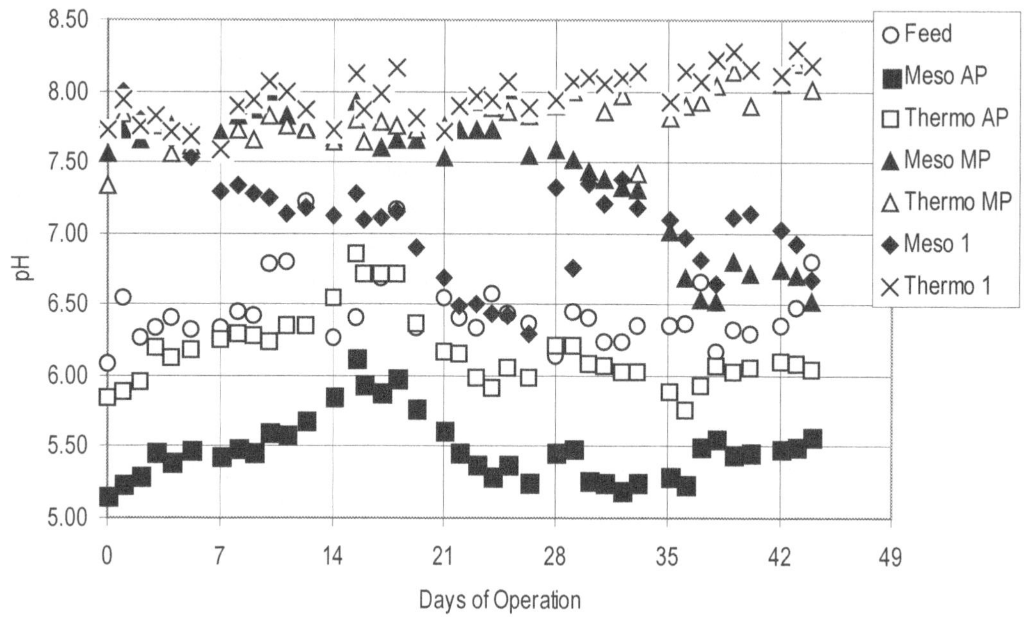

Figure 3-8. pH.

TVA concentrations were approximately 3,000–4,000 mg as acetic acid/L for the acid-phase digesters and, with a few exceptions during periods of instability (Figure 3-9), approximately 1,000–2,000 mg as acetic acid/L for the methane-phase and single-stage digesters. TALK concentrations were more stable with methane-phase and single-stage digesters in the range of 3,000–4,000 mg/L as $CaCO_3$ and acid-phase digesters in the range of 1,500–3,000 mg/L as $CaCO_3$ (Figure 3-10). The TVA/TALK ratio was influenced primarily by the TVA concentrations, and with the exception of the TVA/TALK ratio for the Meso Methane digester (prior to Day 25), the ratios are above levels considered optimum for methane-phase and single-stage operation (i.e., <0.1) (Figure 3-11). High VAs and TVA/TALK ratios in methane-phase and single-stage digesters typically indicate operational problems in mesophilic single-stage digesters.

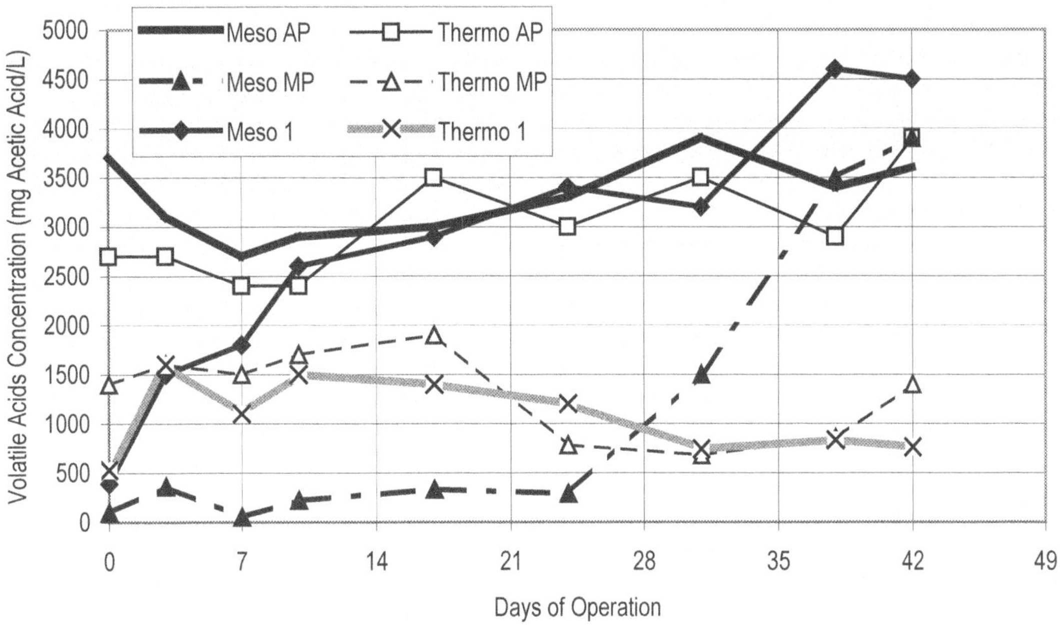

Figure 3-9. Total Volatile Acid Concentration.

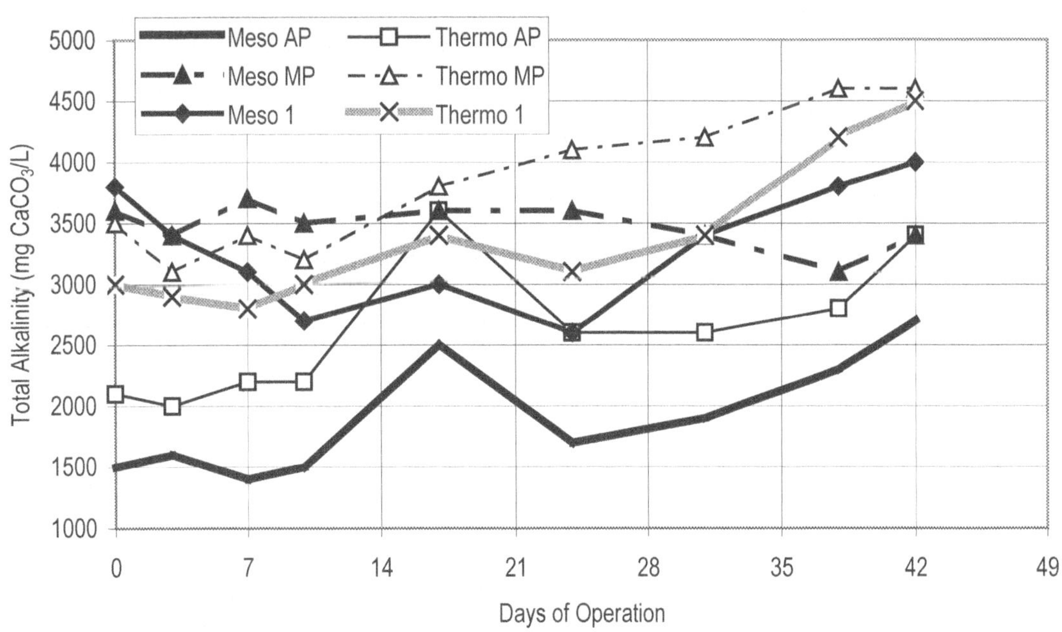

Figure 3-10. Total Alkalinity.

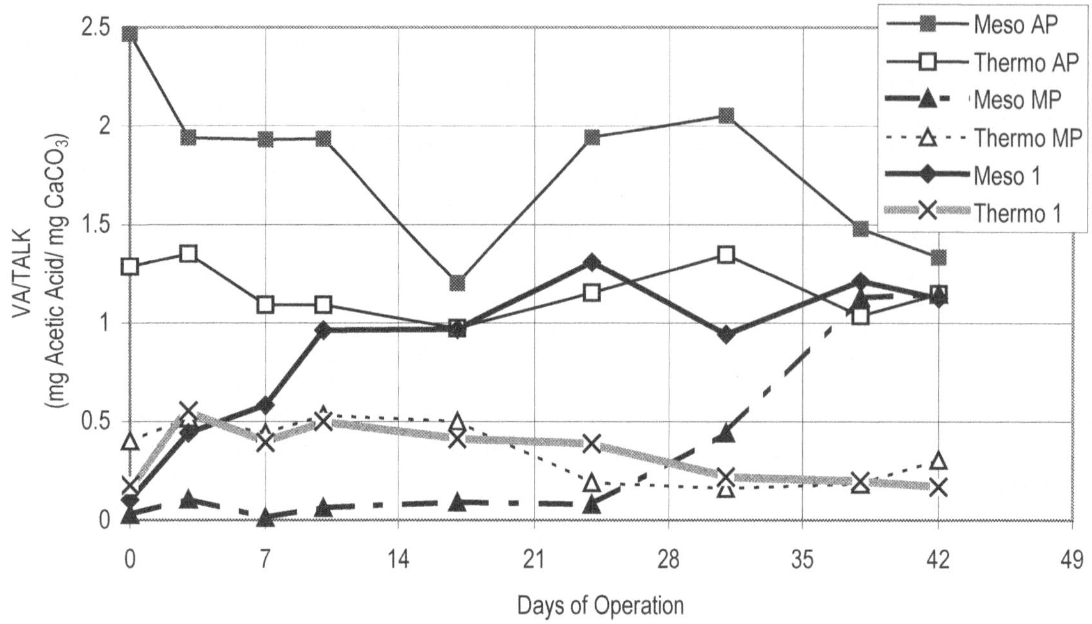

Figure 3-11. Total Volatile Acid/Total Alkalinity (TVA/TALK).

Feed sludge TS concentrations were usually between 3.5% and 5%, although there were a few values as high as 6% to 8% (Figure 3-12). The high feed TS variability may be explained by the pretreatment of the feed sludge. To avoid plugging of lines and pumps, primary sludge was filtered and then allowed to settle overnight then decanted. The filtering, settling, and decanting pretreatment steps may have introduced some variability in the TS concentration and digester loading. Digested sludge TS concentrations are usually approximately 2.5–3% for the methane-phase and single-stage digesters and 3–4% for the acid-phase digesters.

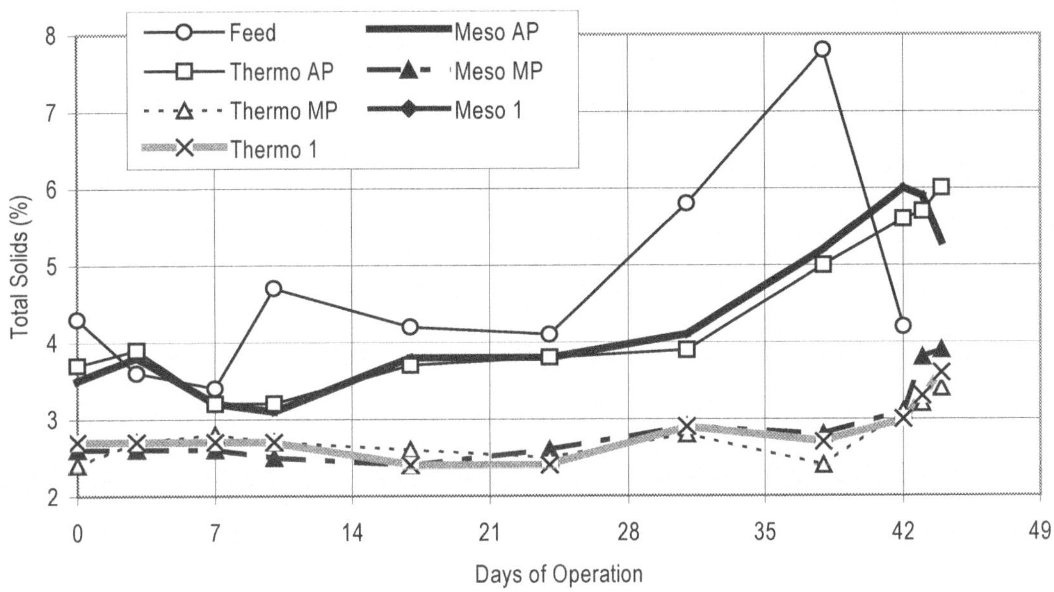

Figure 3-12. Total Solids.

VS concentrations followed patterns similar to those for TS (Figure 3-13). Feed sludge VS concentrations were usually between 60–65% of TS with some lower values (approximately 45%). VS concentrations in the Thermo 1 digester and methane-phase digesters were usually between 50–55% of TS and in the acid-phase digesters approximately 60%. VS reduction (Figure 3-14) was calculated using the Van Kleeck equation and a running average of the feed VS. The running average was taken over an interval equivalent to the digester target HRT (e.g., 15 days for Thermo 1) preceding the collection date for the digested sludge sample. VS reduction for the methane phase digesters included VS reduction occurring in the acid phase (i.e., it represented total reduction by both phases). VS reduction shows substantial variation, however, the VS reduction for the acid-phase digester is usually between 10–20% and between 35–50% for the methane-phase and single-stage digesters. VS reduction in the Meso Control digester declined significantly, from approximately 40% to 20% over the test period.

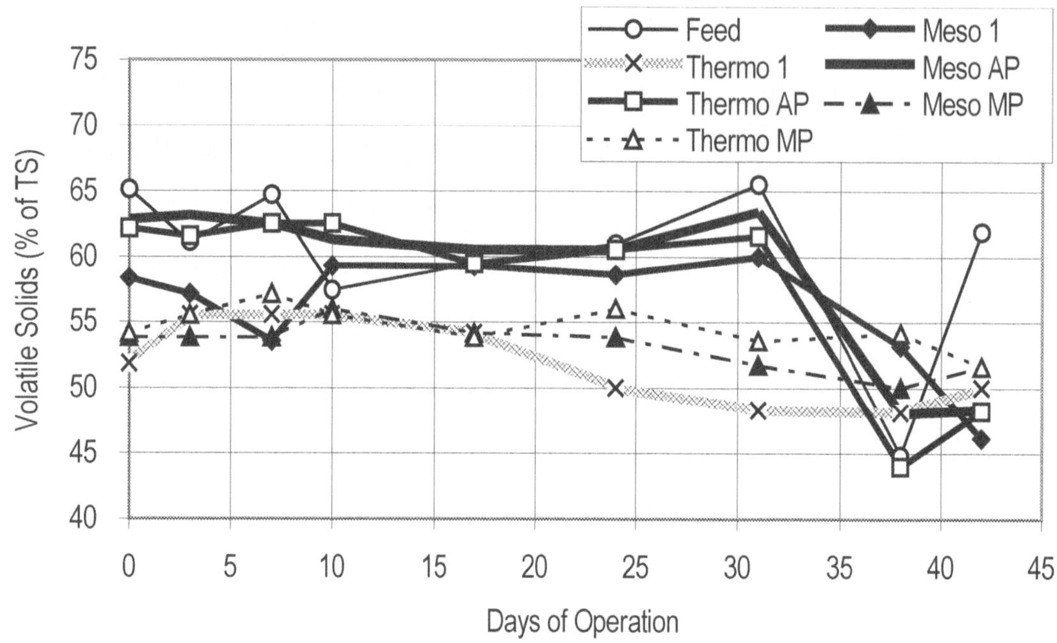

Figure 3-13. Volatile Solids.

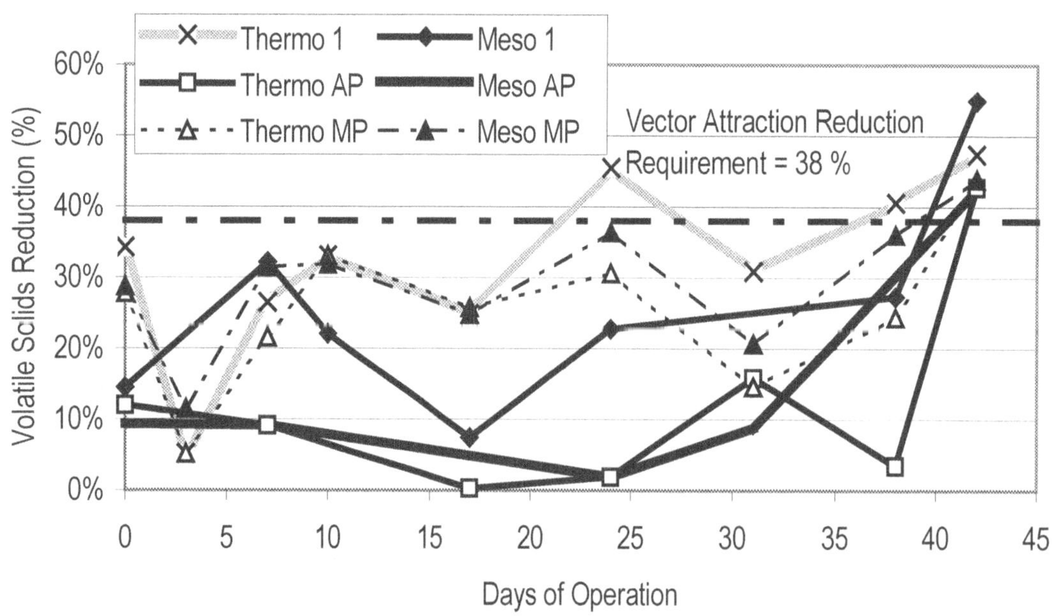

Figure 3-14. Volatile Solids Reduction.

3.4 Effects of Process Parameters on Disinfection

The influence of process parameters on disinfection efficiency was tested through a regression analysis between these parameters and the logarithm of digested sludge (digester effluent, or biosolids) fecal coliform density (log FC_f). Table 3-7 presents the results of this analysis. While R^2 appears to be strong (86.8%), most of the t-statistics are weak ($t < |2.0|$), suggesting that there was little or no relationship between these parameters and log FC_f. The t-values suggest that the parameters with significant influence on log FC_f are temperature (TEMP), digester feed fecal coliform density (log FC_i), thermophilic mean cell residence time (TMCRT), and whether the digester is one- or two-staged. Process parameters without significant influence on log FC_f are pH, ammonia (NH_3), TVA, and TALK.

Table 3-7. Regression Analysis Results 1. Regression Equation: Log FCf = 1.33 + 0.359 pH – 0.123 Temp – 0.00164 NH3 + 0.000226 VA + 0.000429 ALK + 0.619 Log FCi – 0.0963 TMCRT – 0.917 Staged.

Predictor	Coefficient	t-statistic	p-value
Constant	1.331	0.34	0.735
pH	0.3589	0.93	0.357
Temp	-0.12349	-4.43	0.000
NH_3	-0.0016368	-1.87	0.070
TVA	0.0002259	1.82	0.077
TALK	0.0004288	1.64	0.110
Log FCi	0.6194	2.21	0.033
TMCRT	-0.09626	-3.53	0.001
Staged	-0.9168	-3.68	0.001

R^2 = 86.8% Number of data points = 46
F-statistic = 30.29
p-value = 0.000

High TVA concentrations are usually indicative of an upset digester. Throughout Phase 2 testing, all of the digesters experienced some periods of instability marked by high TVA concentrations. Because the regression analysis did not show a significant correlation between TVA and digested sludge (biosolids) disinfection efficiency, it appears that digestion process instability had no impact on the ability of the thermophilic digester to disinfect sludge. The first-stage thermophilic digester also showed significant fecal coliform reductions with high TVA concentrations and with little if any methane gas produced. This suggests that digestion can consistently disinfect feed sludge as long as the adequate temperature, thermophilic MCRT and/or staging are maintained, regardless of whether or not there is a substantial population of healthy methanogenic bacteria.

To improve the regression model, the less significant parameters were removed and the regression analysis rerun. This time the t-values for all parameters were significant (Table 3-8), including the constant—which had a very weak t-value in the previous analysis. For all the parameters considered in the second analysis, t-values improved with the exception of log FC_i, which decreased to a value where it barely met the criterion for significance. The F-statistic rose from 30.29 to 52.10, indicating that the regression model had improved and showed a stronger relationship between all model parameters and log FC_f than in the first analysis.

Table 3-8. Regression Analysis Results 2. Regression Equation: Log FCf = 5.56 – 0.117 Temp + 0.428 Log FCi – 0.0979 TMCRT – 0.789 Staged.

Predictor	Coefficient	t-statistic	p-value
Constant	5.559	3.64	0.001
TEMP	-0.11735	-6.21	0.000
Log FC_i	0.4278	2.08	0.042
TMCRT	-0.09788	-5.84	0.000
Staged	-0.7890	-3.89	0.000

R^2 = 80.3%   Number of data points = 56
F-statistic = 52.10
p-value = 0.000

The resulting model:

$$\text{Log } FC_f = 5.56 - 0.117 \text{ TEMP} + 0.428 \text{ Log } FC_i - 0.0979 \text{ TMCRT} - 0.789 \text{ Staged}$$

Where: ***FC_f*** *is the fecal coliform density in the digested sludge or resulting biosolids*
TEMP *is the digester sludge temperature*
FC_i *is the fecal coliform density in the digester feed sludge*
TMCRT *is the thermophilic MCRT, uses only the volume of thermophilic digester sludge*
Staged *is 1.0 if the digester is staged, 0.0 if not*

was used to predict the log FC_f from 5 digestion systems:

1. single-stage mesophilic digester,
2. thermophilic first-stage (MCRT = 2 days) of a two-stage thermo/meso digester,
3. two-stage thermo/meso digester,
4. single-stage thermophilic digester, and
5. two-stage meso/thermo digester.

All mesophilic digesters were assumed to have a temperature of 35°C, and all thermophilic digesters were assumed to be at 50°C, and log FC_i was assumed to be 6.7, consistent with the average values in Phase 2. The results of this simulation are presented in Table 3-9 and compared with the average values measured during Phase 2.

Table 3-9. Digested Sludge (Biosolids) Fecal Coliform (FC) Densities: Model Predicted vs. Measured in Phase 2. $Log(FC_f) = 5.56 - 0.117\ TEMP + 0.428\ Log(FC_i) - 0.0979\ TMCRT - 0.789\ Staged$.

Digester Configuration	TEMP	TMCRT	Staged	Log FC_f Predicted	Phase 2 Results		
					Log FC_f Mean	Log FC_f Range	No. of Samples
Mesophilic Control	35	0	0	4.33	4.35	3.40-5.20	14
Thermo Acid	50	2	0	2.38	2.18	0.30-4.20	15
Thermo/Meso	50	2	1	1.59	1.97	0.30-2.70	16
Single-Stage Thermo	50	15	0	1.09	1.06	0.30-1.53	13
Meso/Thermo	50	13	1	0.51	0.36	0.30-0.60	14

Based on this empirical model, the single-stage thermophilic digester outperformed the two-staged thermo/meso digester with respect to fecal coliform reduction. The two-staged meso/thermo digester reduced fecal coliforms better than the single-stage thermophilic digester. The model also predicted that most of the fecal coliform reduction was in the first stage thermophilic digester, and less than 1 log was removed in the second stage mesophilic digester. The model predicted that all the digester systems can meet the U.S. EPA Class A fecal coliform requirement (Log $FC_f < 3.0$), except for the single-stage mesophilic digester. The predicted log FC_f values were strongly supported by averaged log FC_f values measured in Phase 2 for all digester configurations considered.

Because aqueous (undissociated) ammonia can be toxic at a high enough concentration, it has often been suggested to be significant by anaerobic digestion for disinfection (Ferran et al., 2002; Pecson and Nelson, 2003; Ward and Ashley, 1978; Yang, 1996). Although the regression analysis described above does not support this suggestion, ammonia may still be contributing to disinfection. Figure 3-15 shows the variation of total ammonia concentration with process configuration over time during this study. The feed sludge has the lowest total ammonia concentration, followed by the mesophilic first stage (2-day MCRT), the thermophilic first stage (2-day MCRT), and then the mesophilic and thermophilic second- and single-stage digesters (13- to 15-day MCRTs).

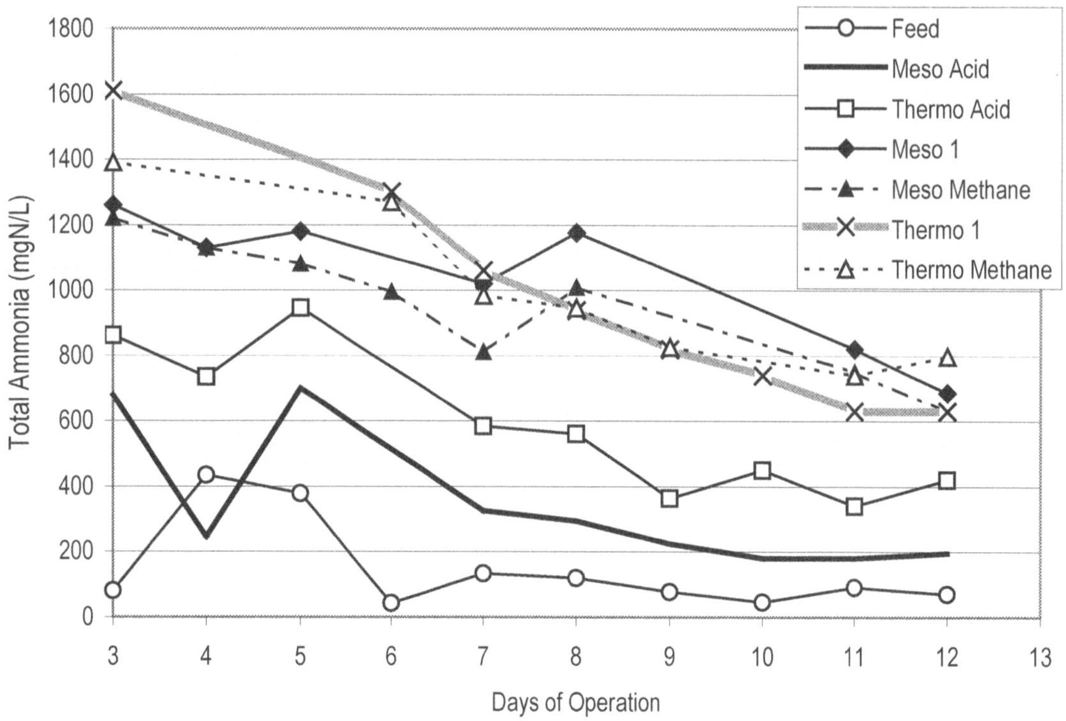

Figure 3-15. Total Ammonia Concentration.

Another regression analysis was performed to quantify the relationship between digester sludge total ammonia concentration (NH_{3out}) and digester sludge temperature (TEMP), total mean cell residence time including both thermophilic and mesophilic sludges), whether the digester was staged or not (Staged = 1 when two-stage, = 0 when single-stage), and digester feed sludge total ammonia concentration (NH_3in). Regression analysis results (Table 3-10) support Figure 3-15. Total MCRT appeared to have the most influence on digester sludge total ammonia concentration, and NH_{3in} concentration was also significant, based on its t-value (> 2.0). The temperature and staged parameters were not shown to be significant based on t-values (< 2.0). Figure 3-15 shows that digester sludge temperature strongly influenced the digester sludge total ammonia concentration, but only at the lower MCRTs.

Table 3-10. Regression Analysis Results 3. The regression equation is: NH3out = -64 + 8.42 Temp + 38.7 MCRT + 58.8 Staged + 0.684 NH3in

Predictor	Coefficient	t-statistic	p-value
Constant	-64.4	-0.31	0.758
Temp	8.423	1.95	0.057
MCRT	38.736	6.52	0.000
Staged	58.83	0.79	0.434
NH3in	0.6838	2.42	0.019

R^2 = 55.3% Number of data points = 54
F-statistic = 15.14
p-value = 0.000

Because more ammonia is generated, probably through the deamination of organic nitrogen in the feed sludge at longer MCRTs, the 2-day MCRT thermophilic digester had lower total ammonia concentrations than the 15-day MCRT mesophilic digester. The thermophilic digester, however, produced better fecal coliform kills than the mesophilic digester even at much lower MCRTs. Temperature is the dominant factor for disinfection, not total ammonia concentration alone; however, higher temperature and pH make a given total ammonia concentration more lethal (Grady et al., 1999). Regression analysis showed that thermophilic MCRT (TMCRT) significantly influenced fecal coliform reduction with higher TMCRTs resulting in higher fecal coliform reductions. This may be due to the longer exposure of fecal coliforms to the higher temperatures, or to higher total ammonia concentrations or both.

3.5 Summary

Acid-phase digester conditions were relatively easy to achieve and maintain. Average TVA concentrations were approximately 3,000 mg as acetic acid /L and gas production was relatively low compared to the methane-phase and single-stage digesters. Stable operation was more difficult to achieve for the single-stage and methane-phase digesters at both mesophilic and thermophilic temperatures. During the study period, each of these digesters experienced instability manifested by higher-than-normal TVA concentrations.

It is difficult to determine the cause(s) of these high TVA conditions. One may have been related to the need to open the digesters for Sentinel Chamber placement or removal. Rubber stoppers sealing the top of the Erlenmeyer flasks (Figure 3-2) were removed periodically for brief periods (1–2 min) to allow chambers to be inserted or removed. A protocol was developed to minimize exposure of digester culture to the atmosphere (and oxygen) that included purging the digester headspace with nitrogen gas. Although, periods of opening and closing digesters do not always correspond to the unstable periods, chambers were replaced and removed from the Mesophilic MP digester starting on October 5, 2000, without any immediate problems. Other digesters, e.g., Meso 1, were never opened for chamber placement, yet experienced similar problems. Feed toxicity could not have been responsible because all digesters received the same feed sludge, and only one or two digesters experienced problems at the same time. Furthermore, problems were observed in individual digesters at different times over the course of the study period.

Because of this instability, it is difficult to compare the process performance of the different digestion processes. Although the effluents from the Thermo 1, Thermo AP/Meso MP, and Meso AP/Thermo MP processes showed VS reductions of less than or very close to the 38% requirement for VAR, it is probable that these processes could achieve higher reductions under more stable conditions. Phase 1 results from this study provided better data for accessing process performance because the larger digesters (30-L) ran for longer periods under more stable conditions (DeLeon et al., 2000).

The three thermophilic digestion processes (Thermo 1, Thermo AP/Meso MP, and Meso AP/Thermo MP) outperformed a control digester (Meso 1) in terms of fecal coliform and pathogen destruction. The three thermophilic processes also showed that they could consistently meet the Class A criterion for fecal coliforms and substantially reduce unusually high *Salmonella* concentrations (4 to 7 log units). Of the three thermophilic processes, only the Meso AP/Thermo MP process and the single-stage Thermo 1 (after the first 11 of operation) consistently reduced both enteric virus and helminth ova densities to the non-detectable levels required of Class A sludge (with one exception on Day 36 for each digester). Pathogen destruction did not appear to be affected by high digester TVA concentrations.

Thermo MP and Thermo 1 digesters demonstrated the highest potential for achieving Class A biosolids on the basis of fecal coliform, *Salmonella*, poliovirus, and helminth ova destruction. These digesters had the longest HRTs at thermophilic temperature (13 and 15 days, respectively, at 50°C). It appears that the Thermo MP performed slightly better than the Thermo 1 digester.

A regression analysis of Phase 2 data showed that thermophilic temperature (TEMP), thermophilic MCRT (TMCRT), feed sludge fecal coliform density, and whether the digester wasone- or two-staged are the most influential factors determining digested sludge fecal coliform density. Regression analysis also confirmed that higher digester MCRTs result in higher digester sludge ammonia and that higher temperature only produced more ammonia at the lower digester MCRTs.

CHAPTER 4.0

PHASE 3 FULL-SCALE PILOT TEST

4.1 Background and Purpose

Bench-scale testing conducted as part of the Screening Evaluation (Chapter 2.0) and Pathogen Destruction Test (Chapter 3.0) demonstrated that two thermophilic anaerobic digestion processes could best achieve Class A standards for biosolids. The two processes, two-phase Meso Acid/Thermo Methane and single-stage Thermo, were considered for full-scale testing. Single-stage Thermo was selected for full-scale testing over a two-stage digester primarily because it was easier to implement at East Bay Municipal Utility District (EBMUD) main wastewater treatment plant (MWWTP) where full-scale single-stage digesters were available with a feed system that could be modified to a fill/ hold/draw mode. Available digester systems with capability for both fill/hold/draw and staging in series were not identified at the time of this pilot test. Further, recent research in thermophilic anaerobic digestion has focused mostly on staged or phased systems that involve more than one digester (Schafer et al., 2003). It was felt that a full-scale pilot study of single-stage thermophilic anaerobic digestion would provide valuable information to wastewater treatment plants limited to single-stage digestion processes.

The primary purpose of the full-scale pilot study was to confirm pathogen destruction results obtained at bench-scale, i.e., that a full-scale single-stage thermophilic anaerobic digester could achieve Class A standards for biosolids. In addition, other important digestion parameters such as process stability, dewaterability, and odor generation would be evaluated and compared to the performance of a single-stage mesophilic anaerobic control digester.

4.2 Materials and Methods

4.2.1 Process

Two full-scale anaerobic digesters at the EBMUD MWWTP were selected for the pilot study and are referred to as the Thermo Test and Meso Control digesters. Operating criteria for these two digesters are shown in Table 4-1.

Table 4-1. Preliminary Operating Criteria.

Item		Thermo Test	Meso Control
Target Operating Temperature	°C	45-50	35
Target Hydraulic Retention Time (HRT)	days	15	15
Feed Sludge		Primary/TWAS mix (approx. 50/50 v/v)	Primary/TWAS mix (approx. 50/50 v/v)
Feeding Regime		Fill/ hold/draw	continuous
Target Loading Rate Range	day	2-3 kg TS/m^3	2-3 kg TS/m^3
Digester Diameter	feet	95	95
Digester Volume	MG	1.5	1.5
Digester Heating		External spiral-type HX	External spiral-type HX
Digester Mixing		Compressed digester gas	Compressed digester gas
Digester Cover Type		Floating	Floating

To assess process performance, the following parameters were measured at the frequencies shown in Table 4-2.

Table 4-2. Process Parameters.

Constituent	Sample	Method	Frequency
Temperature	Digester Sludge	DCS and probes	Continuous
pH	Digester Sludge	SM20; 4500-H+B	Daily
TS	Feed and Digester Sludge	SM20; 2540B	2x/week
VS	Feed and Digester Sludge	SM20; 2540E	2x/week
Fecal Coliforms	Feed and Digester Sludge	SM20; 9221E.1	2x/week
Total Ammonia	Digester Sludge	SM20;	1x/week
TALK	Digester Sludge	SM20; 2320B	2x/week
TVA	Digester Sludge	SM20; 5560C	2x/week
Individual VFAs	Digester Sludge	Gas Chromatography	1x/week
Gas Production	Digester Gas	DCS	Continuous
Gas Composition	Digester Gas	SM20; 2720C	1x/week

The thermophilic anaerobic digester was started-up using a rapid increase in temperature from the mesophilic range (35°C) to the thermophilic temperature (45°C). Feeding was temporarily suspended then started slowly.

The following describes the start-up method for the Thermo Test digester:

1. Float digester cover with process water.
2. Heat process water to the initial thermophilic target temperature of 45°C (113°F).
3. After achieving target temperature, seed with mesophilic sludge from other digesters.
4. Introduce raw feed sludge at a relatively low feed rate (say 1/10 of typical).
5. Monitor TVA, TALK, pH, and gas production routinely.
6. Following the introduction of raw feed sludge, it is expected that TVA concentrations will increase substantially. As the thermophilic methanogen population grows at the increased temperature, TVA concentrations should decrease. As the TVA level approaches more typical concentrations feeding may resume.

7. As the feed rate increases, continuously monitor digester performance on a regular basis. If signs of instability (i.e., increased TVA, reduced gas production) appear, decrease or stop feeding until stability returns. More seed sludge from a stable digester may be added in conjunction with the reduced feeding.
8. Gradually increase the feed rate until the target 15-day HRT is achieved.

The Thermo Test digester was fed intermittently using a fill/hold/draw regime. The fill phase consisted of a series of feed pulses of raw primary and thickened waste-activated sludge that occurred over approximately 18 hours to a designated level. No sludge was withdrawn from the digester throughout the fill period. This method was used because Thermo Test received feed sludge on a rotational basis with the other EBMUD digesters. At the completion of the fill phase, the digester level was maintained with no withdrawal or filling during the hold phase (4–6 hours). The hold phase ensured that all of the feed sludge entering the thermophilic digester was retained by the digester for a minimum period of time, with no opportunity for short-circuiting. The draw phase was relatively rapid (approximately 30 minutes) to a designated sludge level. The phase durations, pumping rates, and digester levels were calculated to achieve a target average 15-day HRT.

Two significant adjustments were made to the operating criteria over the period of the pilot study. The first was to increase the digester temperature on the basis of early pathogen destruction results. Testing (described later) demonstrated that complete helminth ova destruction was not possible at 45°C and within 6 hours, so the operating temperature was increased to 50°C. The heat exchanger for the thermophilic digester was unable to deliver sufficient heat to maintain 50°C at the target feed rate, so the feed rate was reduced from approximately 100,000 gpd to about 43,000 gpd. This translated into an increase in HRT from 15 to 35 days. At the lower feed rate, the digester temperature was maintained in the range of 48°C to 50°C.

4.2.2 Pathogens

The primary purpose of the full-scale pilot study is to demonstrate that the single-stage thermophilic anaerobic digester is capable of producing Class A biosolids. The Class A criteria are:

- Fecal coliforms ≤1,000 MPN/g TS; or
- *Salmonella* < 3 MPN/4 g TS; and
- Enteric virus < 1 plaque-forming unit (PFU)/4 g TS; and
- Helminth ova < 1 viable ovum/4 g TS.

The U.S. EPA Pathogen Equivalency Committee (PEC) also requires that a sufficient number of pathogens be present in the feed sludge to demonstrate adequate destruction (U.S. EPA, 1989). The feed sludge requirements are:

- *Salmonella* > 1,000 MPN/g TSS;
- Enteric virus > 1,000 PFU/g TSS; and
- Helminth ova > 100 viable ova/g TSS.

Most raw wastewater sludges in the U.S. do not contain pathogens at these concentrations, and it is impractical to seed full-scale digesters because of the large number of organisms required, public health concerns, and cost. Sentinel chambers were evaluated during the Phase 2 bench-scale study for this purpose (see Section 3.3.1 for discussion) and found to be a satisfactory method for assessing helminth ova destruction in digestion.

The chambers used for the full-scale tests were similar in design to those used at bench-scale. They were larger and made from different materials. The sentinel chambers used for this test (Figure 4-1) were constructed from 1-in. diameter acrylic tubing with 30-µm stainless steel mesh glued to the 0.25 in. long tube on both ends. *Ascaris* eggs are ovoid and measure 45–70 µm by 35–50 µm (Feacham et al., 1983). The mesh sizes retained helminth ova but not bacteria or viruses.

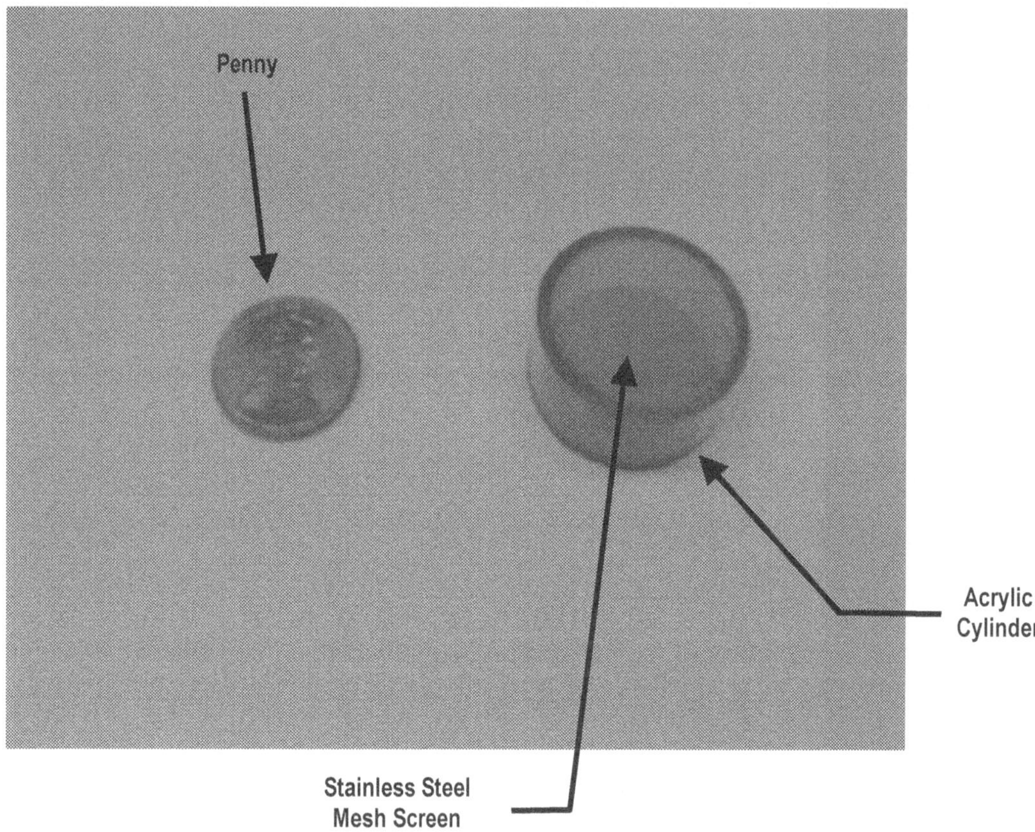

Figure 4-1. Sentinel Chamber.

The chambers were loaded with EBMUD digester sludge, seeded with helminth ova (*A. suum*) then placed in three wire mesh baskets attached to a wire cable at depths approximately 5, 10, and 15 ft (Figure 4-2). At regular time intervals, the cable was pulled out of the digester to retrieve the chambers. At least duplicate samples were obtained for each sample station and period. When triplicate samples were available, one chamber was taken to a second laboratory for confirming analysis. Helminth ova analyses were conducted using Pathogen Equivalency Committee (PEC) approved methods (U.S. EPA, 1999). Control chambers showed survivals ≥94% in both test runs. Temperature sensors were placed at each sample station along the cable to continuously monitor temperatures during the test runs. For the first test conducted at 45°C, chambers were removed at six different times as shown in Table 4-3.

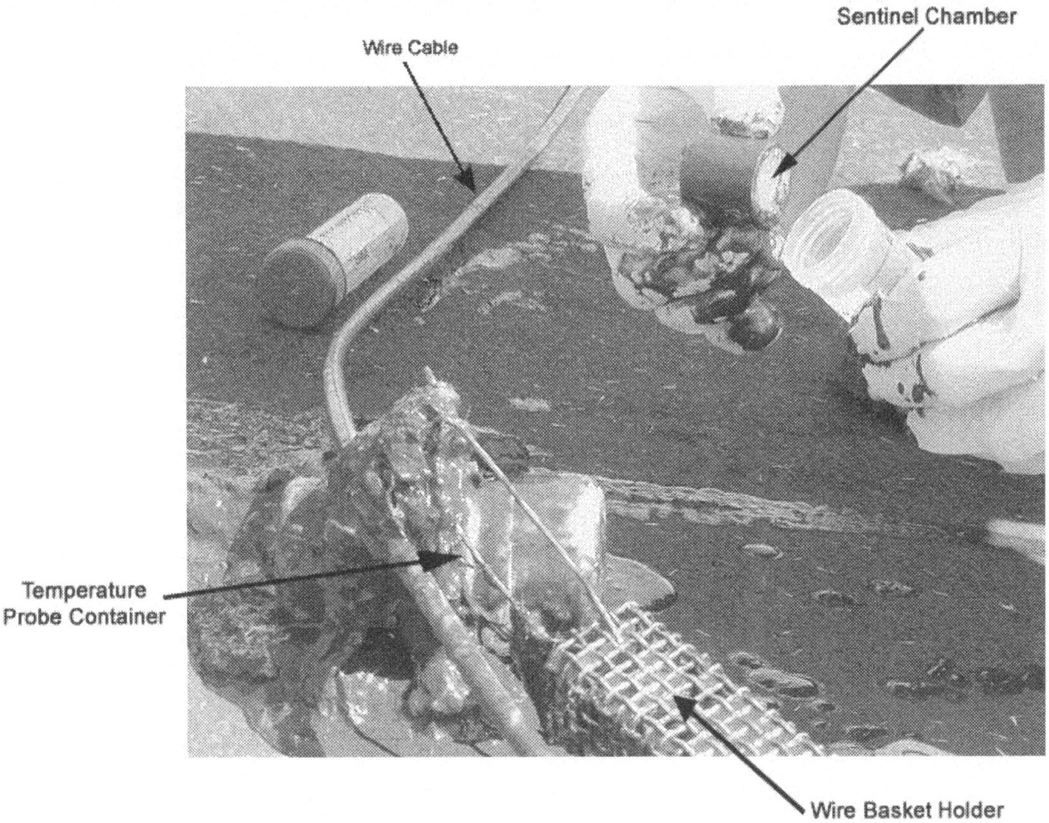

Figure 4-2. In Situ Sentinel Chamber.

Table 4-3. Sentinel Chamber Removal Schedule.

Location in Digester	Removal Times	Replicates	No. Chambers Required
5 feet below surface	2, 4, 6, 8, 10, 12 hours	At all removal times	12
10 feet below surface	2, 4, 6, 8, 10, 12 hours	At all removal times	12
15 feet below surface	2, 4, 6, 8, 10, 12 hours	At all removal times	12

The removal times for the second run were changed to 4, 6, 8, 12, and 24 hours.

At each digester level, a temperature sensor was attached to the sentinel chamber holders to continuously record digester sludge temperature. Submersible data loggers were used to sense and record temperature. Hobo digital dataloggers (StowAway TidbiT) were used with the following specifications:

- Range: -20°C to 50°C (-4 to 122°F)
- Accuracy: ±0.4°C at 20°C (±0.8°F at 70°F)
- Resolution: 0.3°C at 20°C (0.6°F at 70°F)

The goal was to measure viable helminth ova concentration at a specific chamber removal time at all depths. The data would be used to determine the minimum hold time for the fill/hold/draw regime.

Although the full-scale digester cannot be seeded with enteric virus, feed and digested sludge samples (at least six samples of each) were collected from Thermo Test and analyzed for enteric viruses during the course of the pilot study. Enteric virus was analyzed using PEC-approved methods.

4.2.3 Dewatering

Several previous studies (Drury et al., 2001; Garber, 1954; Golueke, 1958; Lee et al., 2001; Rimkus et al., 1982) indicated that thermophilically digested sludges dewater more readily than mesophilically digested sludge. Thermophilic and mesophilic sludges were segregated for dewatering evaluation conducted with full-scale equipment (horizontal solid-bowl type centrifuges) in the field. Concurrently, capillary suction time (CST) tests were conducted in the laboratory.

Typically, digested sludges from the primary digesters at EBMUD are combined in two unheated and unmixed secondary digesters and pumped to the digested sludge wet well in the Dewatering Building. To isolate thermophilically digested sludge for dewatering and analysis, the digested sludge wet well was drained to receive only Thermo Test digested sludge. Digested sludge, cake, and centrate samples were collected for solids capture calculations. To compare with meso digested sludge, this sludge was also isolated in the digested sludge wet well, bypassing secondary digestion/storage. These isolation events were conducted four times over the test period. In addition to the full-scale tests, weekly CST tests were conducted for both Thermo Test and Meso Control digested sludges. The weekly CST test was performed concurrently with the sludge isolation events described above. A summary of dewatering tests is shown in Table 4-4.

Table 4-4. Dewatering Test Schedule.

Analysis	Digester	Sample Type	Frequency
CST (routine)	Thermo Test, Meso Control	Digested sludge	Weekly
Total Solids	Thermo Test, Meso Control	Digested sludge, centrate, cake	Four times during test period
Fecal coliforms	Thermo Test, Meso Control	Cake solids	Four times during test period

4.2.4 Odors

Previous work (Garber, 1975, 1977, 1982; Ghosh et al., 1995; Rimkus et al., 1982; Volpe et al., 1993) has indicated that thermophilic digestion produces a more odorous digested sludge than mesophilic digestion. Possible causes for the increased odor levels include higher concentrations of ammonia, VFAs, and other VOCs in the thermophilic digester gas. EBMUD collected digester gas and digested sludge samples from Meso Control and Thermo Test to quantify the relative odor-formation potential between the two anaerobic digestion processes.

Four digester gas samples each were collected from the Thermo Test and Meso Control digesters from December 30, 2002 to January 16, 2003. Samples were collected in 1-L Tedlar® bags. Each digester gas sample was analyzed by an outside laboratory (Air Toxics) for individual VOCs (modified U.S. EPA method TO-14 using GC/MS full scan) and individual sulfur gases (modified ASTM method D-5504 with GC/SCD). A list of the compounds scanned and the results of these analyses are provided in Appendix B.

Weekly digested sludge liquid stream samples were collected from September 17, 2002 to February 11, 2003. Samples were collected directly from the sludge recirculation lines on each digester and analyzed by the EBMUD Laboratory for individual VFAs.

4.3 Results and Discussion

4.3.1 Process

Standard operating parameters for assessing digester process stability and performance were collected one-two times per week, except for pH (daily) and temperature (continuous). Continuous temperature readings were averaged daily for reporting purposes. Grab samples were collected from the digester recirculation line for laboratory analyses (e.g.,TVA, TALK, TS, and VS). Process parameter averages and standard deviations for the study period (over 5 months) are summarized in Table 4-5.

Table 4-5. Phase 3. Averages and Standard Deviations for Measured Parameters.

Parameter	Units	# of samples		Feed	Thermo Test Period 1[a]	Thermo Test Period 2[b]	Meso Control
Residence time	days	113	average		28	31	45
			st. dev.		6	3	9
			coeff.of var.		0.2	0.1	0.2
Temperature	°C	150	average		46	50	37
			st. dev.		1.1	1.1	2
			coeff.of var.		0.02	0.02	0.06
Methane in gasproduced	%	21	average		51	59	56
			st. dev.		8.7	3.8	6.7
			coeff.of var.		0.2	0.1	0.1
pH		78	average		7.3	7.5	7.3
			st. dev.		0.1	0.1	0.1
			coeff.of var.		0.01	0.02	0.01
Total Alkalinity	mg $CaCO_3$/L	46	average		5200	6000	5700
			st. dev.		400	400	400
			coeff.of var.		0.1	0.1	0.1
Total Volatile Acids	mg acetic acid/L	46	average		200	300	200
			st. dev.		200	200	100
			coeff.of var.		1.0	0.7	0.5
TVA/TALK ratio	mg acetic acid/mg CaCO3	46	average		0.04	0.06	0.03
			st. dev.		0.04	0.04	0.02
			coeff.of var.		0.9	0.7	0.8
Acetic Acid	mg/L	22	average		200	200	100
			st. dev.		100	100	50
			coeff.of var.		0.5	0.5	0.5
Propionic Acid	mg/L	22	average		30	50	40
			st. dev.		20	60	60
			coeff.of var.		0.7	1.2	1.5
Butyric Acid	mg/L	22	average		< MDL	< MDL	< MDL
			st. dev.		NA	NA	NA
			coeff.of var.		NA	NA	NA

Table 4-5. Phase 3. Averages and Standard Deviations for Measured Parameters, continued.

Parameter	Units	# of samples		Feed	Thermo Test Period 1[a]	Thermo Test Period 2[b]	Meso Control
Isobutyric Acid	mg/L	22	average		< MDL	< MDL	< MDL
			st. dev.		NA	NA	NA
			coeff.of var.		NA	NA	NA
Valeric Acid	mg/L	22	average		< MDL	< MDL	< MDL
			st. dev.		NA	NA	NA
			coeff.of var.		NA	NA	NA
Isovaleric Acid	mg/L	22	average		< MDL	< MDL	< MDL
			st. dev.		NA	NA	NA
			coeff.of var.		NA	NA	NA
Total Solids	%	48	average	5.0	2.5	2.5	1.7
			st. dev.	0.9	0.2	0.3	0.2
			coeff.of var.	0.2	0.1	0.1	0.1
Volatile Solids	% of TS	48	average	81	66	62	63
			st. dev.	6	2	2	5
			coeff.of var.	0.07	0.03	0.04	0.08
VS Reduction	%	47	average		56	57	58
			st. dev.		5	6	7
			coeff.of var.		0.1	0.1	0.1
Fecal Coliforms	MPN/g TS	56	average	110,000,000	3,000	10	100,000
			st. dev.	670,000,000	5,000	40	330,000
			coeff.of var.	6.1	1.7	4.0	3.3
Fecal Log Reduction		54	average		4	5	2
			st. dev.		2	1	1
			coeff.of var.		0.4	0.2	0.5

The Thermo Test digester temperature target was 45°C for the first 60 days of the study. During this period, the first sentinel chamber test was conducted and showed insufficient destruction to achieve Class A helminth ova criterion within 4-6 hours (see Section 4.3.2 discussion for pathogen test results). Consequently the target temperature was increased to 50°C, and the target was achieved at about Day 70. For the purposes of evaluating averages and standard deviations for process parameters, the Thermo Test results were separated into two test periods. Period 1 is Day 1 through Day 60 (target temperature was 45°C) and Period 2 was Day 70 through Day 164 (target temperature was 50°C). The target temperature for Meso Control was constant at 35°C throughout the full-scale study.

The average Thermo Test temperature during Period 1 was 46°C with a standard deviation of 1.2°C. Temperatures ranged within 3°C of the target during this period (Figure 4-3). At the start of Period 2, Thermo Test temperatures were close to the target of 50°C for about 20 days. Then winter storms and colder ambient temperatures increased the radiant heat loss from the digester making it difficult to maintain a stable operating temperature around the target. Thermo Test temperature decreased to almost 46°C at one point during this period, but was usually ≥48°C. The average Thermo Test temperature was 50°C during Period 2 and showed a standard deviation of 1.1°C. The Meso Control digester temperature ranged between 35° and 40°C, and was usually above the target (Figure 4-4). The average Meso Control temperature is 37°C with a standard deviation of 2.0°C.

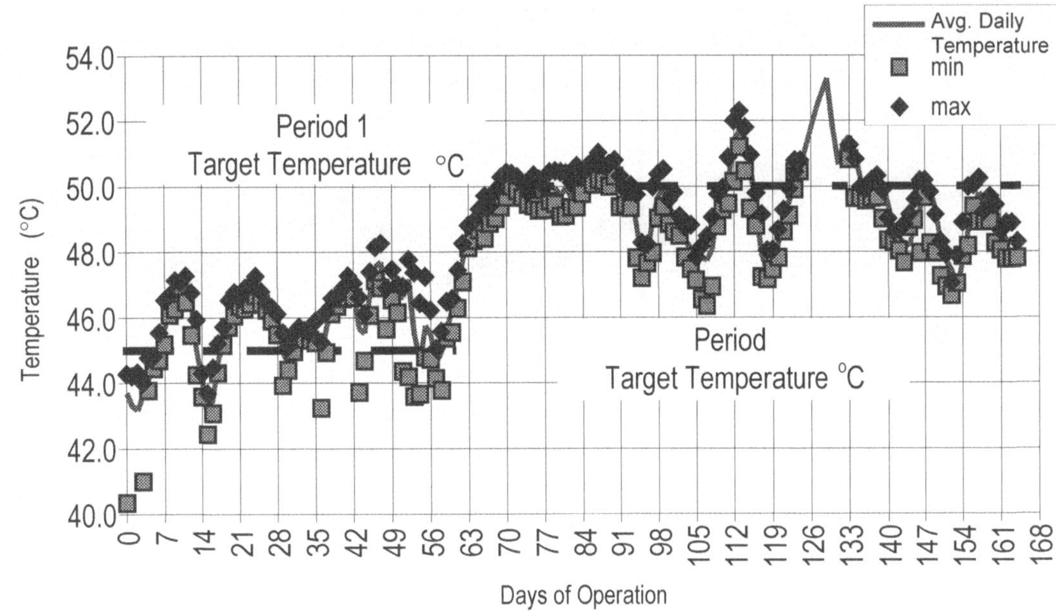

Figure 4-3. Thermo Test Temperature.

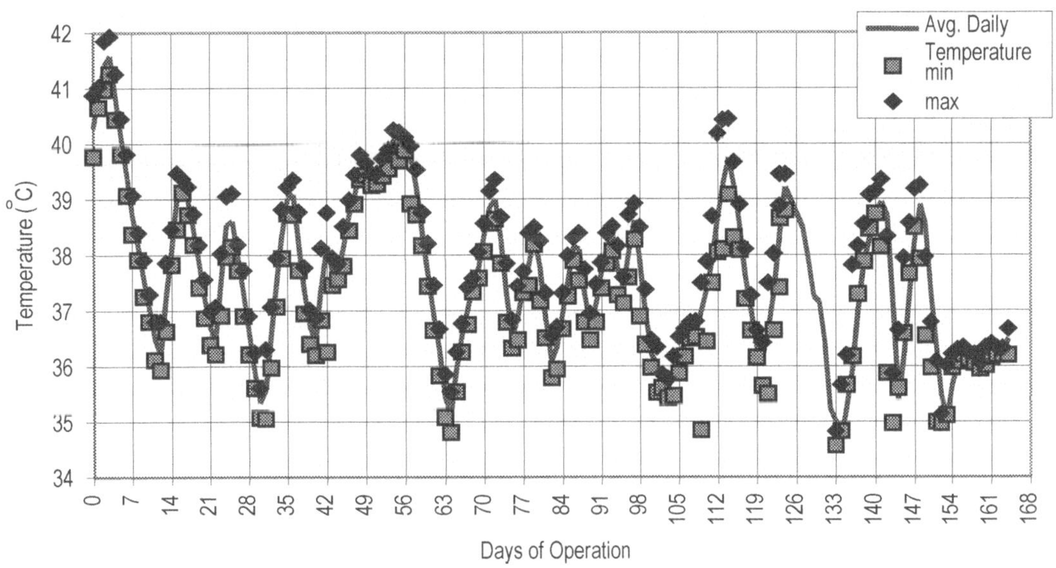

Figure 4-4. Meso Control Temperature.

The Thermo test digester had an average HRT of approximately 30 days throughout the study. The Meso Control average HRT is substantially higher at approximately 45 days. A 30-day running average HRT for each digester is shown in Figure 4-5.

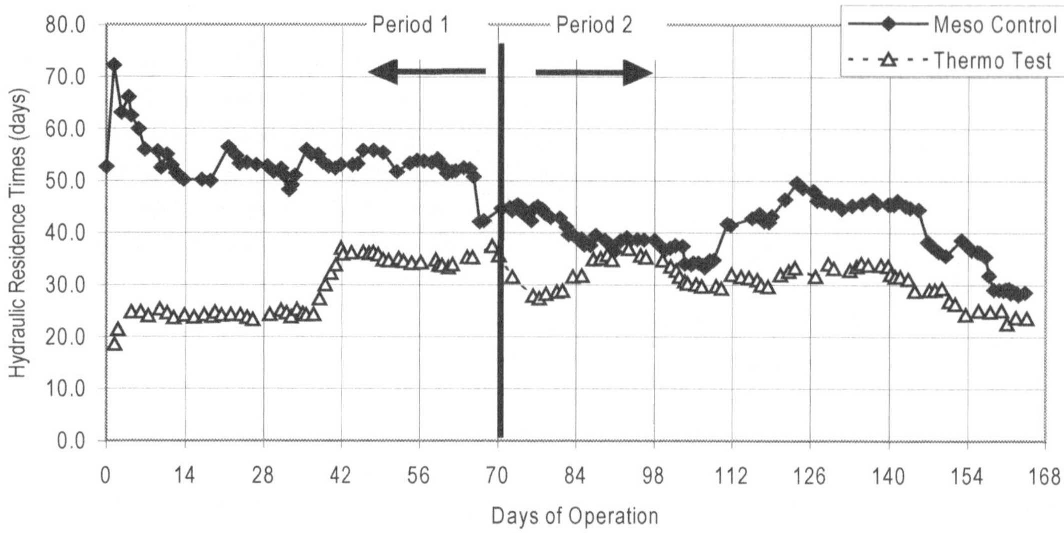

Figure 4-5. Hydraulic Residence Times (30 day running average).

The average pH values for the Meso Control and Thermo Test digesters were both approximately 7.3 during Period 1 (Figure 4-6). They diverged as the Thermo Test temperature increased during Period 2, when it averaged approximately 7.5. A similar trend was observed for TALK, which increased with Thermo Test operating temperature (Figure 4-7). During Period 1, the average TALK for Thermo Test was 5,200 mg/L as $CaCO_3$, and this average increased to 6,000 mg/L as $CaCO_3$ during Period 2. The Meso Control average TALK concentration was 5,700 mg/L as $CaCO_3$. Bench-scale data for single-stage digesters (Table 2-2) showed no significant differences in pH or TALK at different operating temperatures. At about Day 70, the Meso Control HRT decreased from approximately 55 to 40 days, while Thermo Test HRT varied less and Thermo Test HRT became similar to Meso Control HRT.

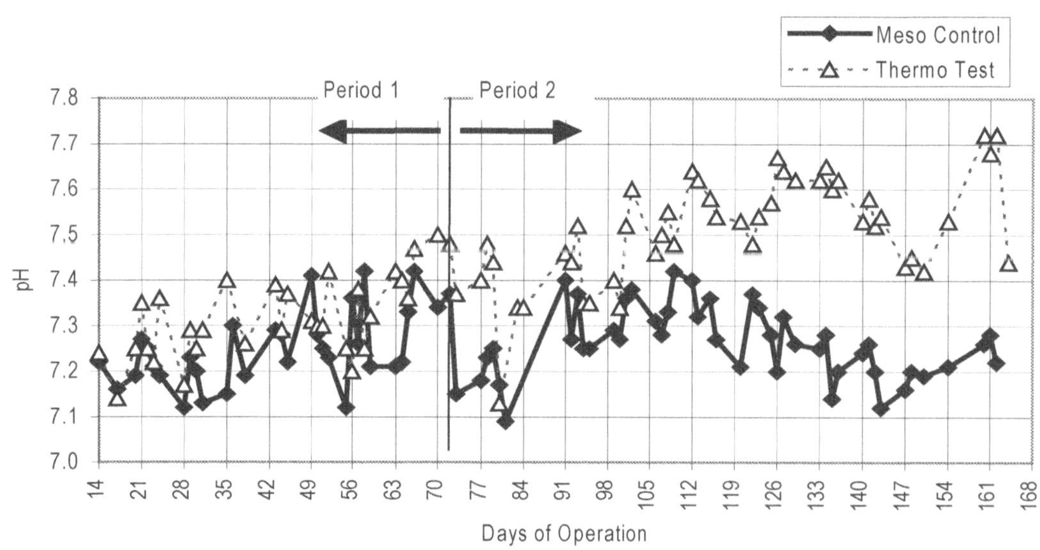

Figure 4-6. pH.

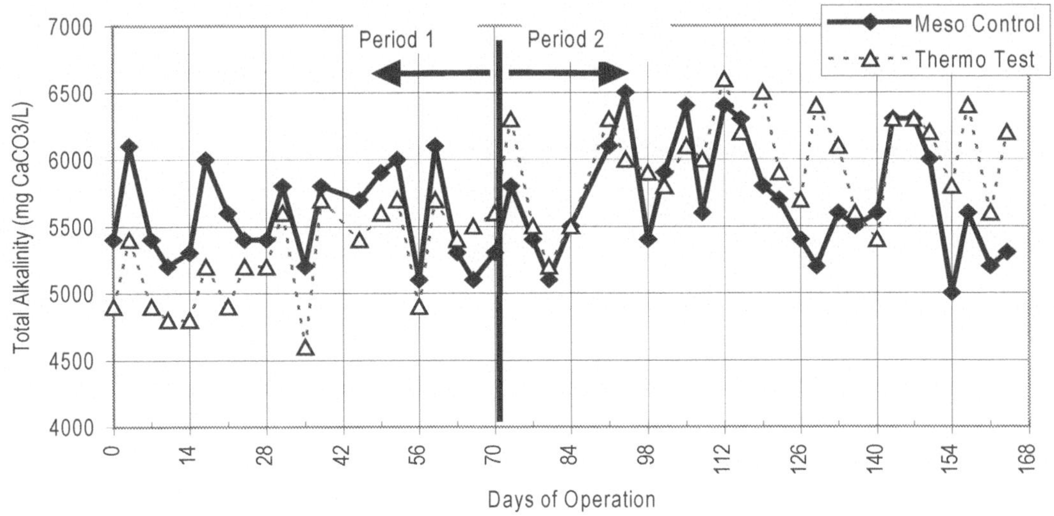

Figure 4-7. Total Alkalinity.

TVA concentrations were usually less than 200 mg/L as acetic acid for both Thermo Test and Meso Control digesters during Period 1 with the exception of two spikes (about 430 mg/L as acetic acid for Meso Control and 1,000 mg/L as acetic acid for Thermo Test) (Figure 4-8). During Period 2, Thermo Test showed several spikes above 800 mg/L as acetic acid. Meso Control and Thermo Test Period 1 average TVAs were 200 mg/L as acetic acid, and the Thermo Test average increased to 300 mg/L as acetic acid during Period 2. During the bench-scale testing, higher and more fluctuating behavior of TVA concentrations was observed at higher thermophilic temperatures. TVA/TALK ratio for Thermo Test was usually less than the target maximum of 0.10 (Figure 4-9), and the maximum value was about 0.20. TVA/TALK ratio for Meso Control was generally lower than that for Thermo Test, and with one exception was less than 0.10.

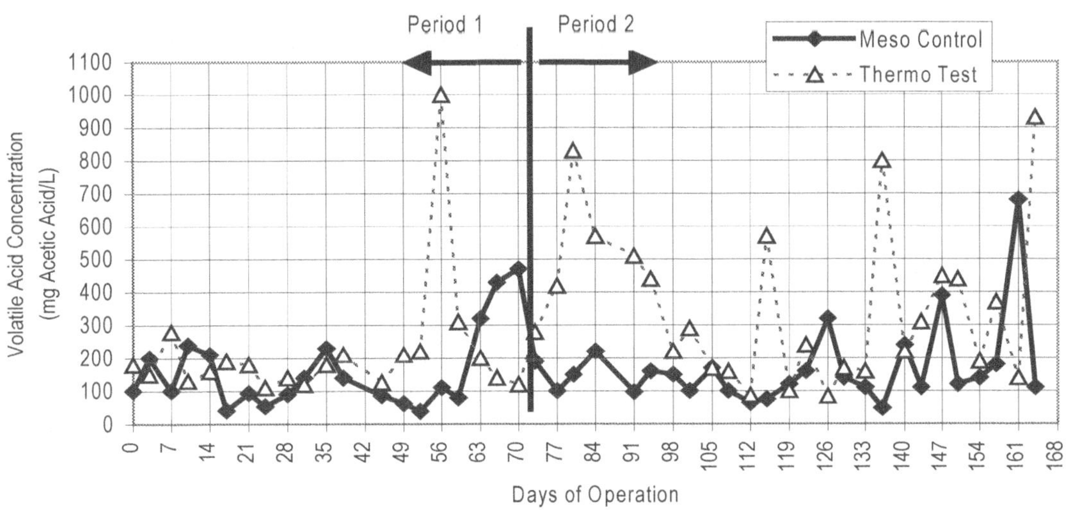

Figure 4-8. Total Volatile Acids.

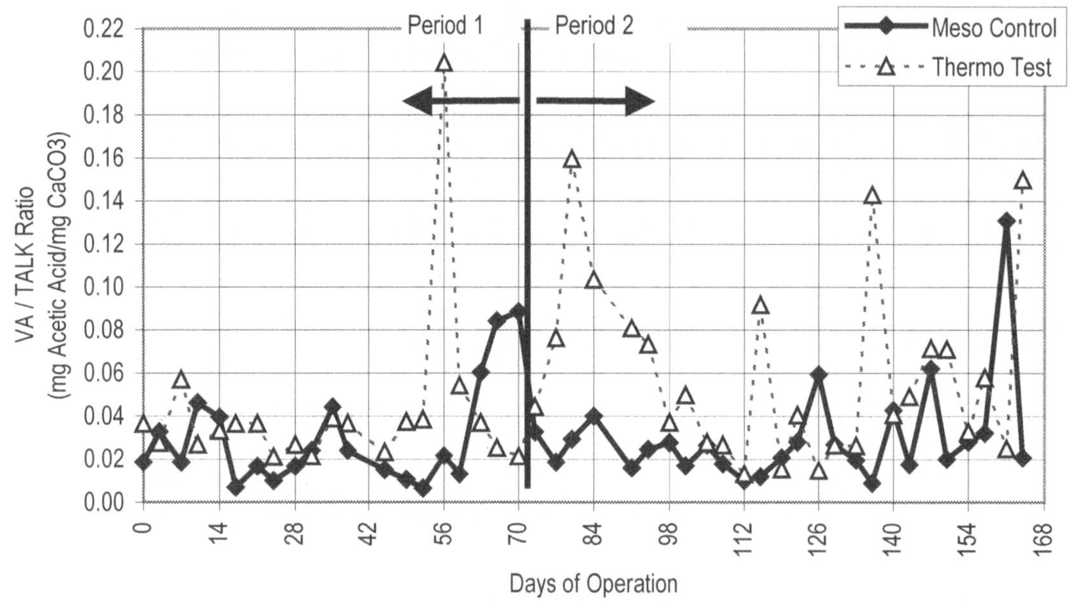

Figure 4-9. Total Volatile Acid/Total Alkalinity Ratio (TVA/TALK).

TS and VS concentrations for Meso Control and Thermo Test digesters were similar and relatively constant (Figures 4-10 and 4-11). There was a significant drop in VS concentrations for both digesters and the feed between Day 105 and Day 112. The cause for this sudden decrease is not certain but may be related to large stormwater flows. VSR was calculated using the Van Kleeck equation. The influent VS used in the calculation was the average value from a period equal to the digester target HRT prior to the date the effluent VS sample was collected. VSR generally ranged between 50% and 65% for both digesters, with no significant difference in performance (Figure 4-12). Both digesters performed consistently above U.S. EPA's 38% requirement for VAR. The decrease in VS concentrations at approximately Day 105 resulted in a corresponding increase in VS reduction followed by a decrease in the reduction rate, but otherwise the reduction rate was constant throughout the test.

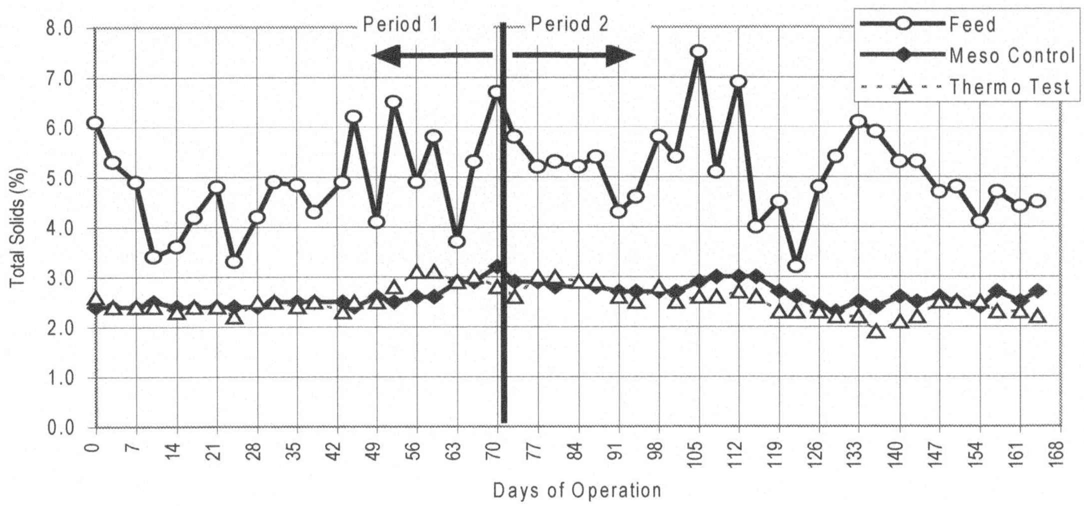

Figure 4-10. Total Solids.

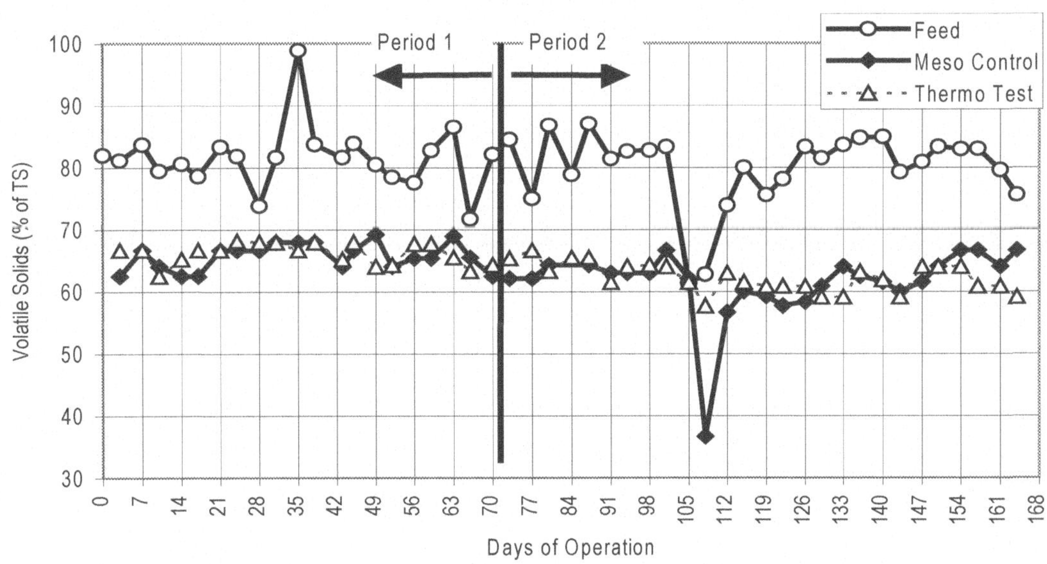

Figure 4-11. Volatile Solids.

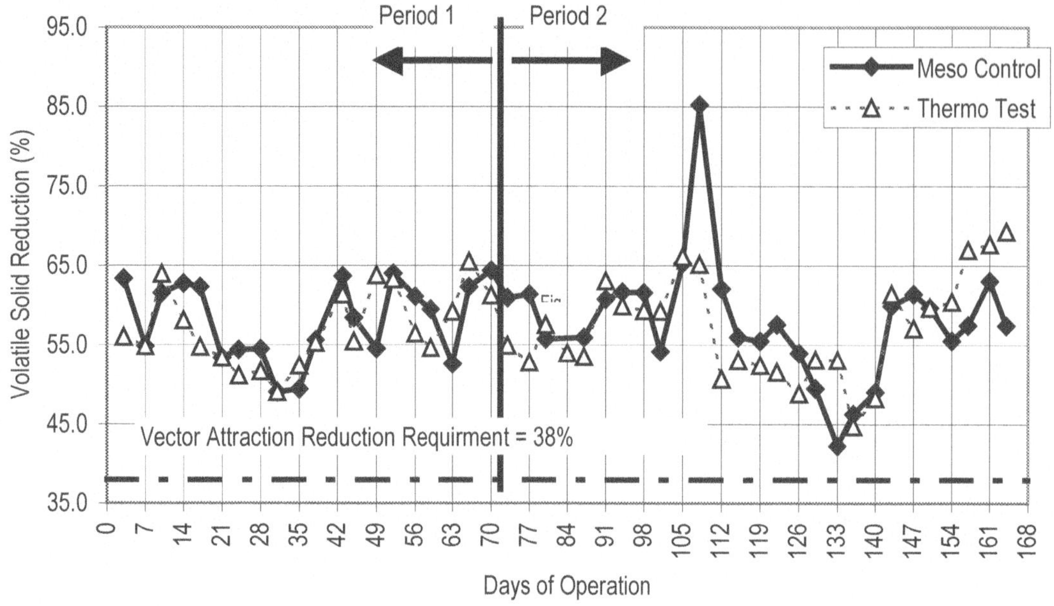

Figure 4-12. Volatile Solids Reduction.

The total ammonia concentrations for Meso Control and Thermo Test digesters were similar and relatively constant throughout the test, ranging between 1,200 and 1,650 mg NH_3-N/L (Figure 4-13). Averages of specific VFA concentrations for the two digesters (Meso Control vs. Thermo Test Period 2) showed higher concentrations of acetic and propionic acid (55 mg/L vs. 37 mg/L) in the Thermo Test digester compared to the Meso Test digester (Figure 4-14). The higher operating temperature appears to result in a higher propionic acid concentration, which may cause odors.

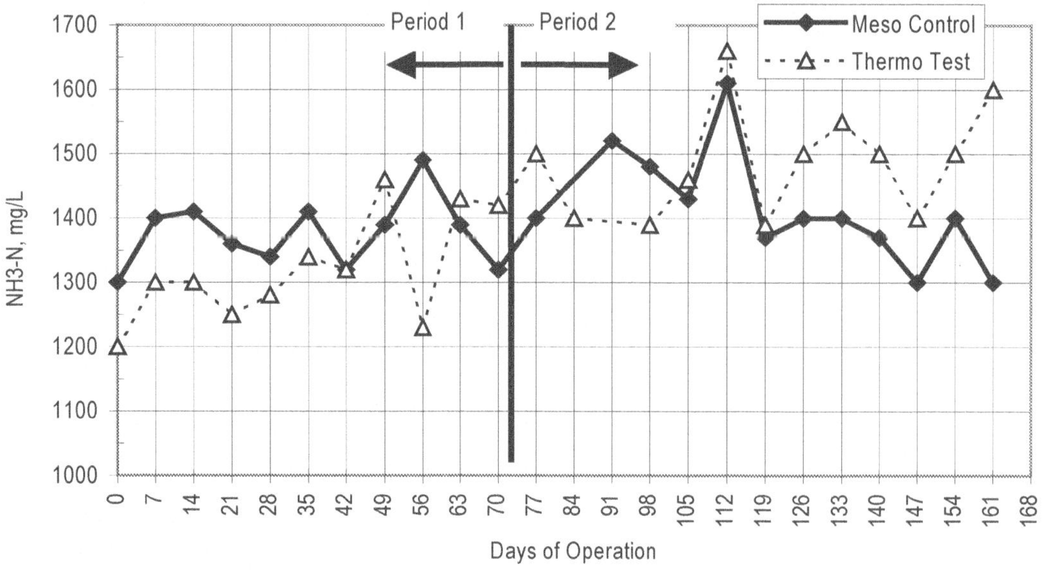

Figure 4-13. Total Ammonia Concentrations.

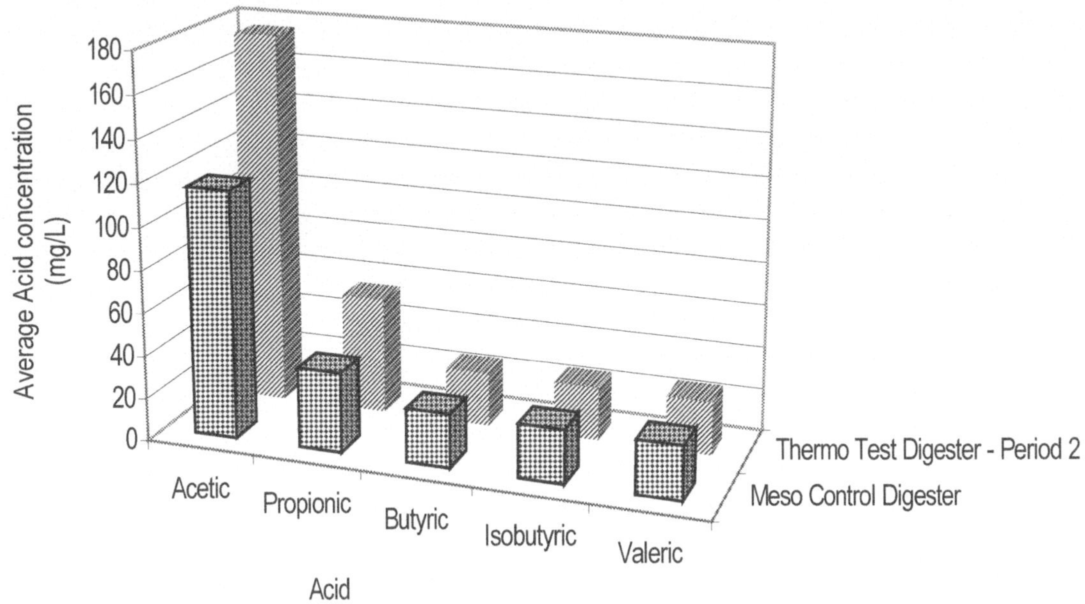

Figure 4-14. Average Concentrations of Specific VFAs.

Digester methane gas concentration was constant throughout the test period for both digesters, and there was no significant difference between Meso Control and Thermo Test (Figure 4-15). Both digesters had gas compositions in a range of 50–65% methane, (usually about 60%), values that are typical for anaerobic digesters treating municipal wastewater sludges.

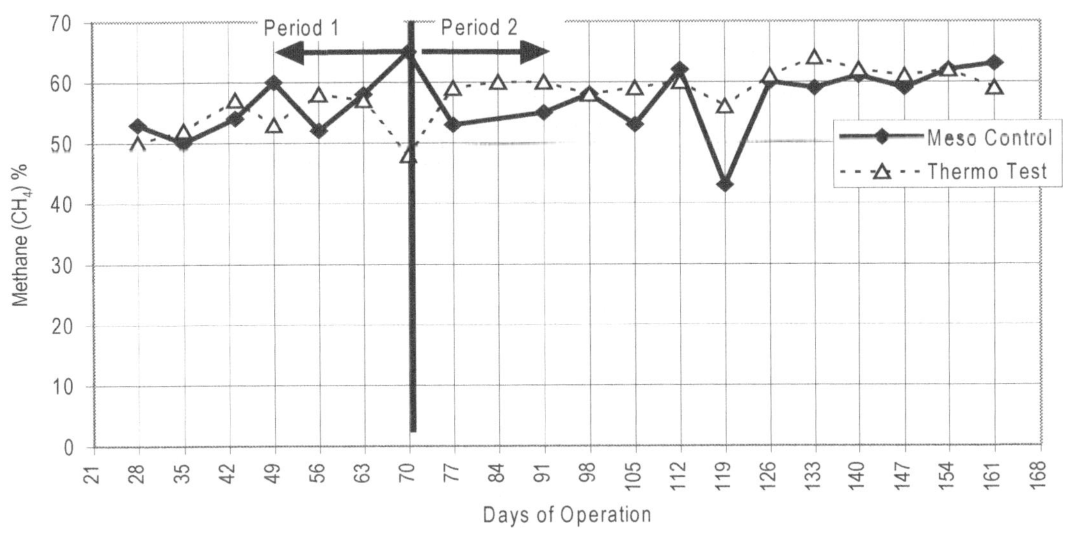

Figure 4-15. Gas Composition.

A lithium chloride tracer test was conducted to determine the mixing characteristics of the Thermo Test digester on October 23, 2002. A lithium chloride slug was injected into the digester to achieve a target concentration of 2.75 mg/L lithium. Nine samples were collected about 2 hours following injection to determine tracer dispersal (Figure 4-16). The sample concentrations throughout the digester ranged from 2.2 mg/L to 2.9 mg/L with a mean of 2.5 mg/L and a standard deviation of 0.3 mg/L, indicating that the tracer slug was well dispersed throughout the digester.

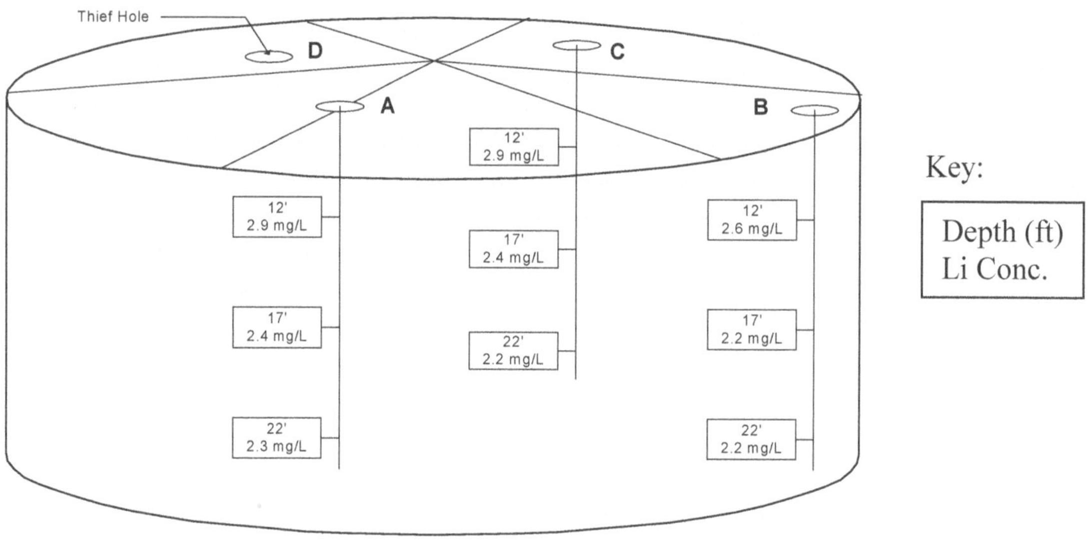

Figure 4-16. LiCl Trace Study—Tracer Dispersal (2 Hours Following Slug Introduction of LiCl).

Digester samples were analyzed for lithium over a 90-day period (approximately three detention times) to develop a washout curve (Figure 4-17). The measured concentrations compared well to the theoretically predicted concentrations indicating that the digester behaved like a continuous stirred tank reactor (CSTR). The washout curve was linearized on a log scale, and a regression analysis yielded an R^2 value of about 94% (Figure 4-18), further supporting characterization of the Thermo Test digester as a CSTR.

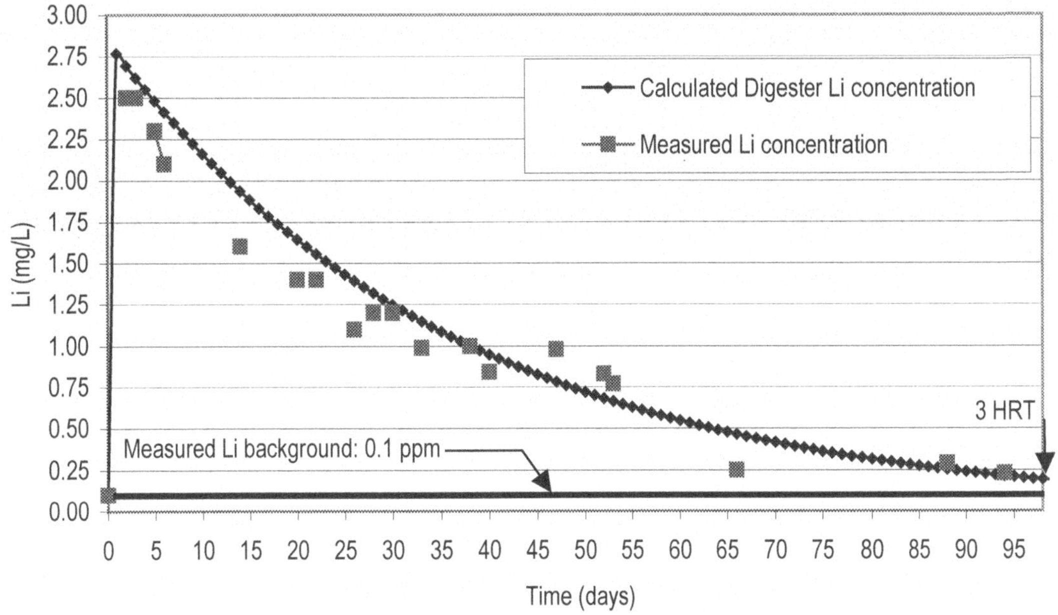

Figure 4-17. Lithium Tracer Test — Theoretical and Measured Washout Curves.

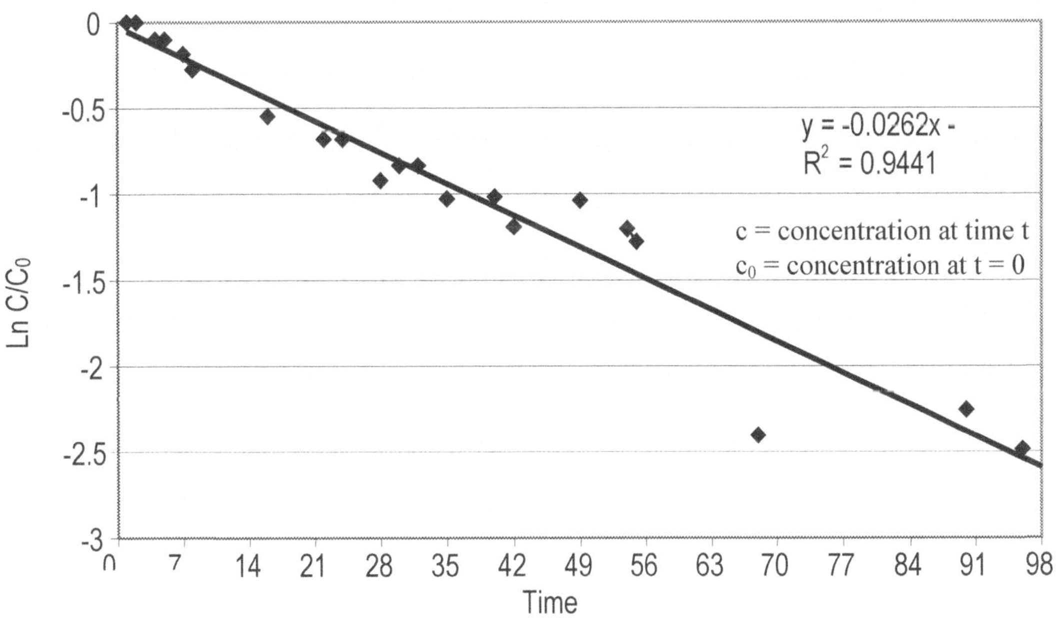

Figure 4-18. Linearized Tracer Washout Curve.

4.3.2 Fecal Coliforms and Pathogens

Thermophilic digester fecal coliform results were most variable early in the test but became fairly constant after Day 70 (Figure 4-19). This was due primarily to a change in sampling protocol and possibly also to an increase in the Thermo Test temperature from 45° to 50°C. At the beginning of the study, fecal coliform samples were collected from the digester recirculation line without regard to the fill/hold/draw schedule. It is possible that some of these samples were taken shortly after feeding when fecal coliform concentrations would be relatively high. After Day 70, fecal coliform samples were collected only after the hold period to ensure that the sample would be representative of digested sludge effluent. Following the change in sampling protocol, thermo digester fecal coliform concentrations consistently met the Class A criterion (<1,000 MPN/g TS) and often were undetectable. Meso Control digester results rarely met the Class A criterion despite the relatively low fecal coliform densities in the feed sludge (approximately 10^6 MPN/g TS).

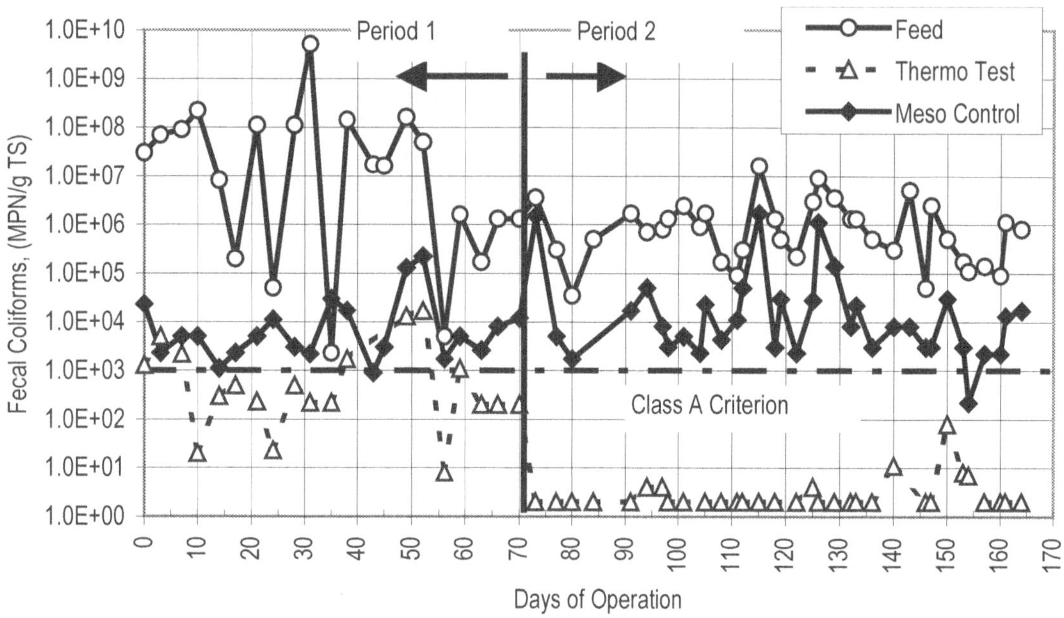

Figure 4-19. Fecal Coliforms.

Helminth ova survival was assessed using sentinel chambers in two test runs. The results from the first sentinel chamber test conducted at an operating temperature of 45°C showed substantial survival, even after 12 hours in the thermophilic digester (Table 4-6 and Figure 4-20). The lowest ova viability observed was 75% in the bottom chamber after 12 hours, and this value was substantially higher than the Class A criterion of 0% viability. Consequently, the digester temperature was increased to 50°C and a second test was conducted (Table 4-7). Following this only one test chamber showed any viable helminth ova. This chamber was for a bottom sample taken after 12 hours in the digester. It had 10 viable ova out of 1,178 counted (viability = 0.85%). Because all of the other 32 chambers contained no viable ova (especially those retrieved preceding the 12-hour chamber at 4, 6, and 8 hours) the finding of this "positive" chamber was puzzling. Two explanations are offered:

- The chamber containing viable helminths was collected at 4 hours and mislabeled. The samples were divided into two shipments for analysis at the Excelsior-Sentinel Lab, Ithaca, NY. Although extreme care was taken to properly label samples during preparation for shipment, mislabeling may have occurred.

- The chamber containing viable helminths was contaminated by the control samples. To avoid this possibility, control samples were analyzed after all of the other samples. However, it was noted that the chamber containing viable helminths was processed just prior to the controls.

Table 4-6. Sentinel Chamber Test No. 1 at 45°C. All Units in % Viable Helminth Ova.

Sample Time (hours)	Digester Sample Position[1]	Chamber No. 1		Chamber No. 2	
		% viable	No. of Ova counted	% viable	No. of Ova counted
Controls		94	148	97	142
		99	134	98	124
		99	106		
2	Top	97	140	97	190
	Middle	94	156	96	148
	Bottom	96	121	96	158
4	Top	96	177	97	113
	Middle	95	120	94	192
	Bottom	94	116	97	144
6	Top	94	110	96	157
	Middle	93	136	96	118
	Bottom	95	129	94	168
8	Top	94	139	94	112
	Middle	92	160	96	167
	Bottom	96	163	93	143
10	Top	94	108	96	143
	Middle	89	142	92	168
	Bottom	90	134	91	110
12	Top	88	160	89	161
	Middle	80	166	81	166
	Bottom	77	144	75	114

[1] Indicates position in digester that chamber was tested. Top, middle, and bottom are approximately 5, 10 and 15' below the liquid surface of the digester.

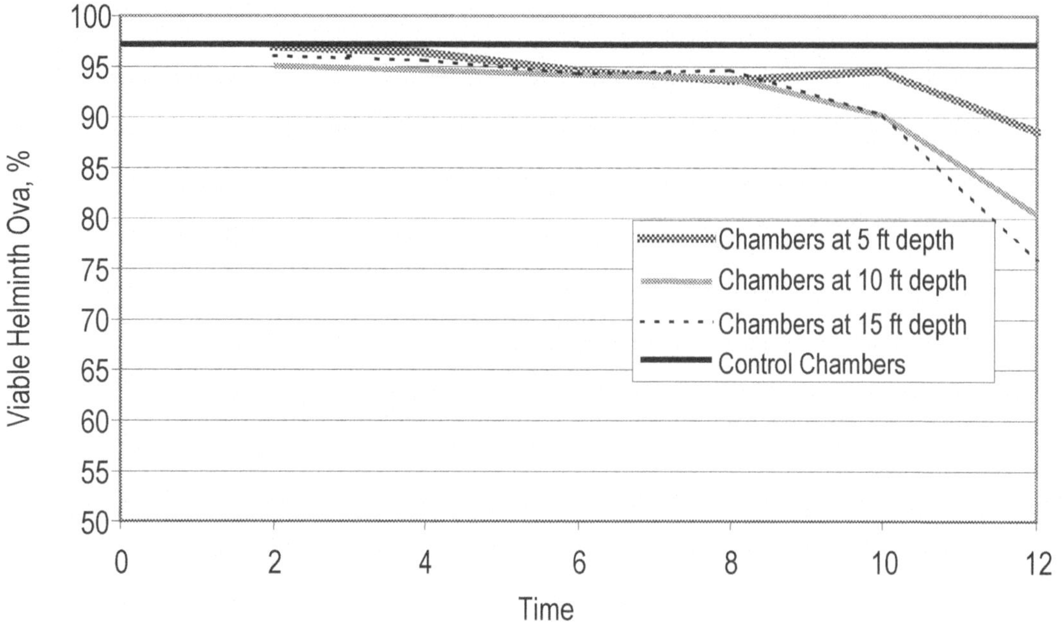

Figure 4-20. Sentinel Chamber Test No. 1 at 45°C.

Table 4-7. Sentinel Chamber Test No. 2 at 50°C. All Units in % Viable Helminth Ova.

Sample Time (hours)	Digester Sample Position[1]	Chamber No. 1		Chamber No. 2		Chamber No.3[2]
		% viable	# of Ova counted	% viable	# of Ova counted	% viable
Controls		95	1307	NA	NA	NA
		99	1584	NA	NA	NA
		96	1200	NA	NA	NA
4	Top	0	1033	0	1000	NA
	Middle	0	1055	0	1064	0
	Bottom	0	1038	0	1240	NA
6	Top	0	1010	0	1014	NA
	Middle	0	1013	0	1015	0
	Bottom	0	1090	0	1055	NA
8	Top	0	1012	0	1072	NA
	Middle	0	1105	0	1000	0
	Bottom	0	1004	0	1040	NA
12	Top	0	1003	0	1101	NA
	Middle	0	1014	0	1063	NA
	Bottom	0.85	1178	0	1018	NA
24	Top	0	1022	0	1023	NA
	Middle	0	1098	0	1012	0
	Bottom	0	1111	0	1123	NA

[1] Indicates position in digester that chamber was tested.
Top, middle and bottom are approximately 5, 10, and 15' below the liquid surface of the digester.

[2] Triplicate sample analyzed by BioVir Laboratories, Inc.
All other samples analyzed by Excelsior - Sentinel

Although the survival of a small number of helminth ova in the 12-hour chamber cannot be explained, the project team is confident that these results demonstrate that the Class A helminth ova criterion can be achieved in a semi-continuously fed thermophilic digester at 50° C with a minimum 4-hour holding period.

During the second test run, recorded temperatures at each sampling station were slightly higher than the 50° C target temperature within a range of 50–52°C (Figure 4-21). The distributed control system (DCS) temperature was measured by RTD probe near the 15-ft digester depth and recorded continuously. The temperatures measured by DCS were slightly less than those measured by the probes but never less than 50°C.

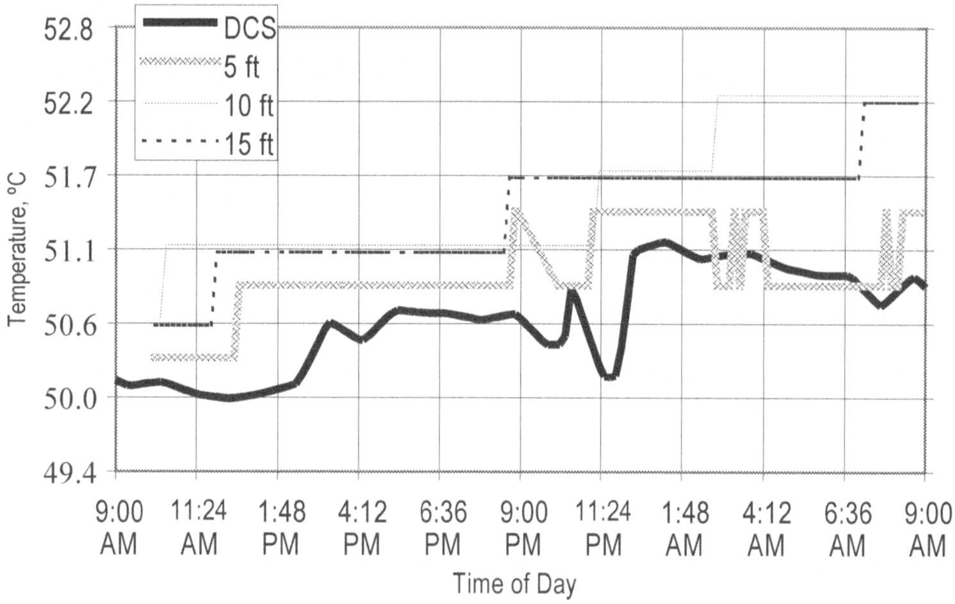

Figure 4-21. Sentinel Chamber Test No. 2 Temperatures: DCS and Data logger Measurements at Various Depths.

Other pathogen testing included analyzing feed and digested sludges for helminths and enteric virus (Table 4-8). Because extensive testing was conducted to assess helminth ova destruction using the sentinel chambers, only the digester feed sludge was tested for helminths. Three feed sludge samples showed undetectable or very low concentrations (1 ovum/4 g TS) of helminth ova. The same feed sludge samples showed very low levels of enteric virus, ranging from 2–20 PFU/4 g TS. Thermo Test samples obtained on the same dates showed non-detectable levels of enteric virus. These results confirm that there were insufficient concentrations of helminths and enteric virus in EBMUD feed sludge to adequately demonstrate destruction by observing background levels of these organisms.

Table 4-8. Thermo Test Digester Pathogen Concentrations at 50°C.

Helminth Ova Results
(No. viable helminth ova / 4 g TS)

Days of Operation	Feed Sludge
105	<1
111	<1
119	1

Enteric Virus Results
(PFU / 4 g TS)

Days of Operation	Feed Sludge	Thermo Digested Sludge
105	2	<1
111	5	<1
119	20	<1

4.3.3 Dewatering

Dewatering test data included sludge temperature, polymer dose, CST, TS, and fecal coliform concentration (cake solids only) (Table 4-9). There were 4 test events and 2 samples from each digester were taken at different times for each test event.

Table 4-9. Dewatering Test Results.

		Meso Control						
	Days of Operation	Temp °C	Feed Flow gpm	Poly Flow gpm	Polymer %	CST s	TS %	Fecal MPN/g TS
Digested Sludge	133	49	133	12.5	9.4	990	2.2	NA
	133	48	132	12.3	9.3	997	2.4	NA
	140	46	150	15.2	10.1	1004	2.2	NA
	140	45	148	15.5	10.5	957	2.2	NA
	147	44	147	17.0	11.6	1171	1.9	NA
	147	45	153	17.3	11.3	897	1.9	NA
	161	44	130	12.3	9.5	1173	2.4	NA
	161	45	128	12.7	9.9	1006	2.2	NA
Digested Sludge w/ Polymer	133	46	NA	NA	NA	18	1.3	NA
	133	44	NA	NA	NA	17	1.4	NA
	140	43	NA	NA	NA	21	1.1	NA
	140	42	NA	NA	NA	20	1.4	NA
	147	41	NA	NA	NA	11	1.8	NA
	147	42	NA	NA	NA	8	1.7	NA
	161	41	NA	NA	NA	13	1.2	NA
	161	42	NA	NA	NA	16	1.9	NA
Centrate	133	45	NA	NA	NA	14	0.4	NA
	133	45	NA	NA	NA	13	0.4	NA
	140	43	NA	NA	NA	17	0.4	NA
	140	43	NA	NA	NA	18	0.4	NA
	147	40	NA	NA	NA	10	0.3	NA
	147	41	NA	NA	NA	11	0.3	NA
	161	29	NA	NA	NA	10	0.2	NA
	161	28	NA	NA	NA	10	0.2	NA
Cake	133	NA	NA	NA	NA	NA	26.0	50
	133	NA	NA	NA	NA	NA	27.0	30
	140	NA	NA	NA	NA	NA	23.0	280
	140	NA	NA	NA	NA	NA	23.0	170
	147	NA	NA	NA	NA	NA	25.0	1700
	147	NA	NA	NA	NA	NA	24.0	900
	161	NA	NA	NA	NA	NA	23.0	700
	161	NA	NA	NA	NA	NA	24.0	130

Table 4-9. Dewatering Test Results, continued.

		Meso Control						
	Days of Operation	Temp °C	Feed Flow gpm	Poly Flow gpm	Polymer %	CST s	TS %	Fecal MPN/g TS
Digested Sludge	133	32	131	12.4	9.5	578	2.5	NA
	133	32	130	12.4	9.5	639	2.4	NA
	140	36	150	15.9	10.6	476	2.4	NA
	140	35	154	16.0	10.4	351	2.4	NA
	147	34	150	16.8	11.2	567	2.5	NA
	147	35	147	16.4	11.2	600	2.4	NA
	161	34	125	12.6	10.1	468	2.5	NA
	161	34	125	11.9	9.5	425	2.4	NA
Digested Sludge w/ Polymer	133	31	NA	NA	NA	7	2.5	NA
	133	32	NA	NA	NA	7	2.2	NA
	140	35	NA	NA	NA	12	1.4	NA
	140	34	NA	NA	NA	13	1.2	NA
	147	33	NA	NA	NA	9	0.5	NA
	147	34	NA	NA	NA	10	2.6	NA
	161	32	NA	NA	NA	14	2.8	NA
	161	33	NA	NA	NA	12	2.4	NA
Centrate	133	32	NA	NA	NA	9	0.3	NA
	133	31	NA	NA	NA	10	0.5	NA
	140	34	NA	NA	NA	15	0.7	NA
	140	34	NA	NA	NA	14	0.7	NA
	147	28	NA	NA	NA	8	n/a	NA
	147	33	NA	NA	NA	12	0.9	NA
	161	32	NA	NA	NA	17	0.9	NA
	161	32	NA	NA	NA	13	0.8	NA
Cake	133	NA	NA	NA	NA	NA	22.0	30,000
	133	NA	NA	NA	NA	NA	22.0	13,000
	140	NA	NA	NA	NA	NA	20.0	14,000
	140	NA	NA	NA	NA	NA	20.0	17,000
	147	NA	NA	NA	NA	NA	22.0	14,000
	147	NA	NA	NA	NA	NA	21.0	2,300
	161	NA	NA	NA	NA	NA	20.0	50,000
	161	NA	NA	NA	NA	NA	20.0	22,000

Cake TS data shows that the thermophilic digested sludge dewatered to a higher TS concentration than the mesophilic digested sludge (Figure 4-22). Thermo Test cake TS concentrations ranged from 23% to 27% (average 24.4%) compared to Meso Control concentrations ranging from 20% to 22% (average 20.9%). Polymer doses were relatively constant, ranging from 9.5% to 11.6%. There does not appear to be a correlation between the small differences in polymer doses and cake TS concentrations.

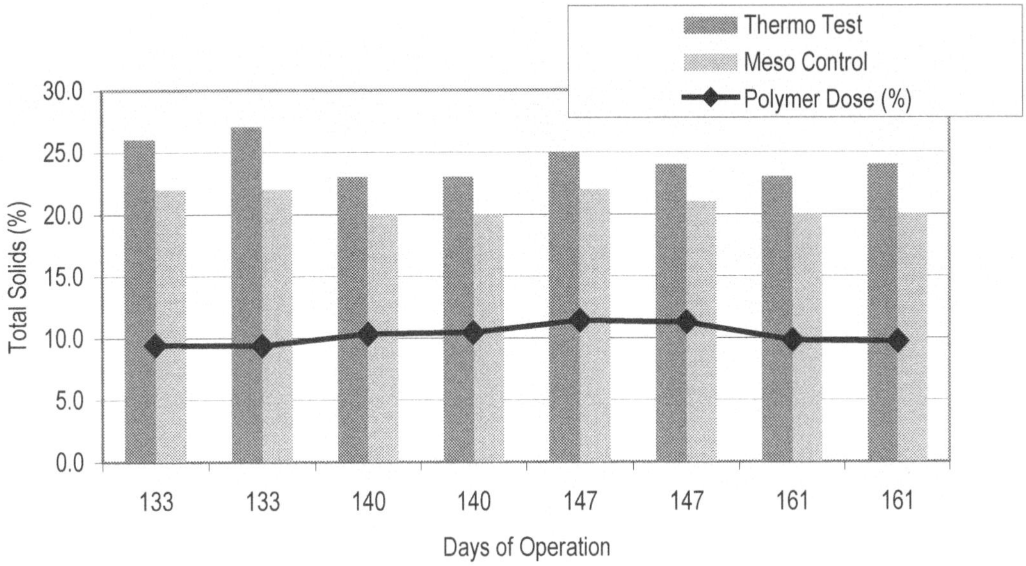

Figure 4-22. Cake Total Solids.

The CST test data directly contradict the findings of the cake TS data, indicating that Meso Control sludge should dewater more easily (Figure 4-23). CST is inversely related to dewaterability, with lower CST values typically indicating greater potential for dewaterability and higher cake TS concentrations. Meso Control sludge CSTs ranged between 425 and 638 s, (average 513 s) compared to Thermo Test sludge CSTs, which ranged between 897 and 1173 s (average 1,024 s). Adding polymer to the digested sludge samples reduced the CST by 1 to 2 orders of magnitude. Meso Control sludge with polymer CSTs ranged between 6 and 14 s (average 10.5 s) compared to Thermo Test sludge with polymer CSTs that ranged between 8 to 21 s (average 15.5 s). The CST test was conducted with digester sludge samples that were heated to approximately the ambient digester temperature, i.e., Meso Control samples were at 35°C, and Thermo Test were at 50°C, to more closely resemble the sludge temperatures in the dewatering centrifuge. Sludge temperature affects dewatering potential because cooler sludges are more viscous and more difficult to dewater.

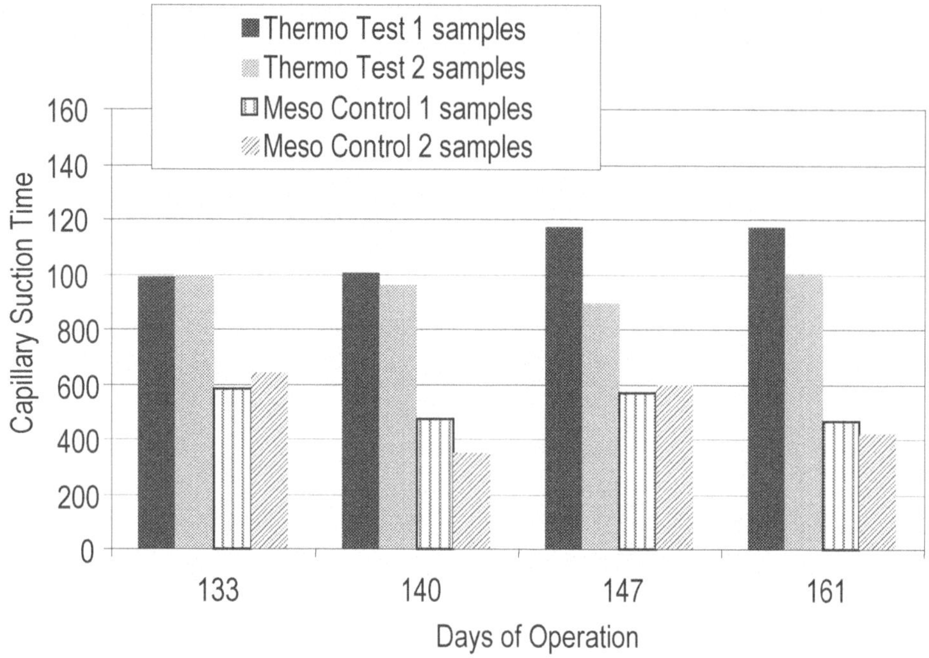

Figure 4-23. Capillary Suction Time (CST).

With one exception, solids capture efficiency for each test event was consistently higher for Thermo Test digested sludge than for the Meso Control (Table 4-10, Figure 4-24). Thermo Test capture efficiencies ranged from 84% to 94% (average 87.7%) compared to Meso Control capture efficiencies, which ranged from 66% to88% (average 74.2%). Fecal coliform concentrations in the cake solids were highest in the Meso Control cake, ranging from 2,300 to 50,000 MPN/g TS with a geometric mean of approximately 15,500 MPN/g TS (Figure 4-25). Thermo Test cake fecal coliform concentrations ranged from 30–1,700 MPN/g TS with a geometric mean of about 240 MPN/g TS. Onlyone of the Thermo Test samples exceeded the Class A fecal coliform criterion of 1,000 MPN/g TS.

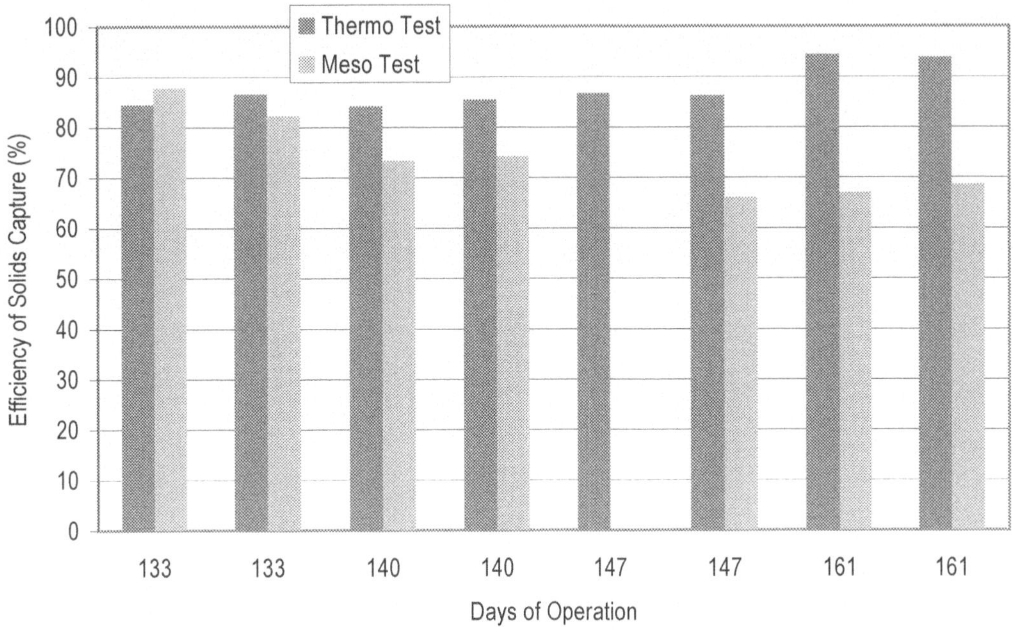

Figure 4-24. Efficiency of Solids Capture.

Table 4-10. Efficiency of Solids Capture.

		Digested Sludge TS Conc. (%)	Centrate TS Conc. (%)	Cake TS Conc. (%)	Solids Capture Efficiency	Average Efficiency
	Symbol	Cs	Cr	Cc	(%)	(%)
	Equation	Efficiency = [1-Cr(Cc-Cs)/Cs(Cc-Cr)]*100%				
Thermo Test	133	2.2	0.4	26.0	84.4	87.7
	133	2.4	0.4	27.0	86.5	
	140	2.2	0.4	23.0	84.1	
	140	2.2	0.4	23.0	85.4	
	147	1.9	0.3	25.0	86.7	
	147	1.9	0.3	24.0	86.3	
	161	2.4	0.2	23.0	94.4	
	161	2.2	0.2	24.0	93.8	
Meso Control	133	2.5	0.3	22.0	87.8	74.2
	133	2.4	0.5	22.0	82.2	
	140	2.4	0.7	20.0	73.4	
	140	2.4	0.7	20.0	74.2	
	147	2.5	n/a	22.0		
	147	2.4	0.9	21.0	66.1	
	161	2.5	0.9	20.0	67.0	
	161	2.4	0.8	20.0	68.6	

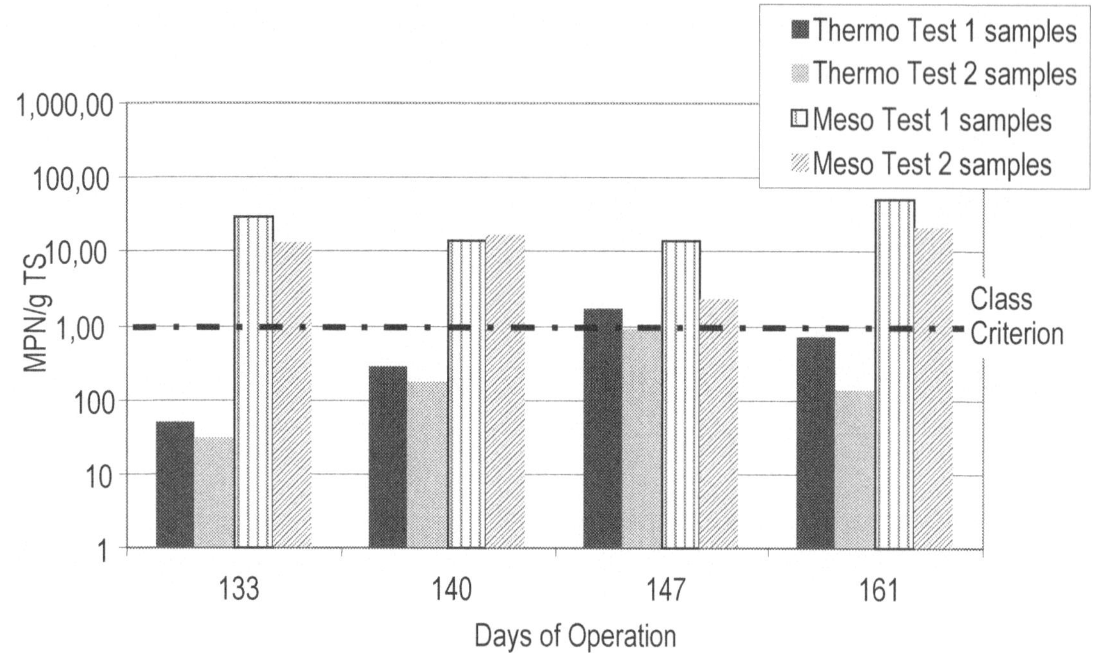

Figure 4-25. Cake Fecal Coliform.

In summary, the Thermo Test sludge dewatered more readily than the Meso Control sludge, as shown by the higher cake TS and greater TS capture efficiency. Thermo Test also produced cake with fecal coliform concentrations that were almost 2 logs less than Meso Control and nearly always less than the Class A criterion. The dewatering results were not supported by the results of the CST test, a surrogate indicator of dewaterability. Although CST results indicated that Meso Control digested sludge should dewater more readily than Thermo Test, actual field results indicated the contrary. The CST test may not be a suitable indicator of dewaterability when comparing mesophilic and thermophilic sludges.

The dewatering test conditions described here were not typical of standard EBMUD operations. As noted in Section 4.2.3, primary digester sludges at EBMUD are ordinarily combined in a secondary digester, unmixed and unheated. Under typical operating conditions, mesophilic and thermophilic sludges are cooled to ambient temperatures prior to dewatering. Cooler sludges are more viscous, and it is expected that cake solids concentrations for both thermophilic and mesophilic digested sludges might be lower than those shown during this test had they been stored in a secondary digester. However, bypassing the secondary digesters was necessary to isolate the sludges for this test.

4.3.4 Odors

Odor potential was evaluated by analyzing digester gas samples for VOCs and sulfur compounds. The digested sludge was also analyzed for individual VFAs. Digester gas results for Meso Control and Thermo Test are presented in Table 4-11. Thermo Test showed two to three times the concentration of total VOCs as the Meso Control (Figure 4-26). The H_2S concentrations in gas were higher in Thermo Test on three of four sampling events, (Figure 4-27). However, no additional reduced sulfur compounds were detected in the digester gas samples, except for carbonyl sulfide in the thermophilic sample on December 30, 2002. The individual VOC and sulfur compound concentration results for each digester gas sample are included in Appendix B (Tables B-1 and B-2, respectively).

Table 4-11. Summary of Digester Gas Total VOC and H_2S Results.

Days of Operation	Total VOCs (ppbv)		H_2S (ppm)	
	Meso	Thermo	Meso	Thermo
118	1,193	3,990	180	370
126	1,357	4,002	350	400
133	1,105	4,002	530	420
135	3,907	6,178	690	860

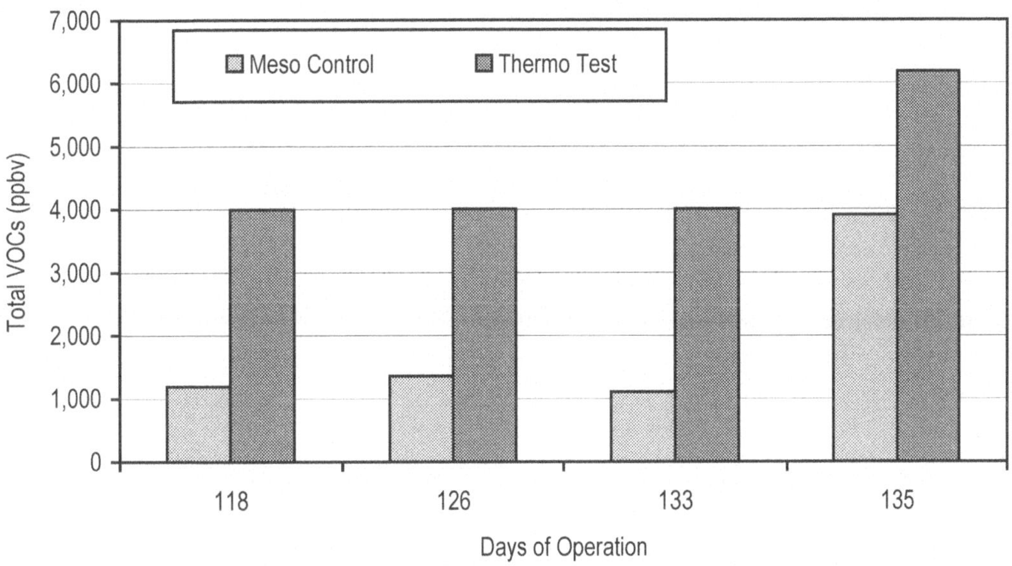

Figure 4-26. Digester Gas Total Volatile Organic Compounds.

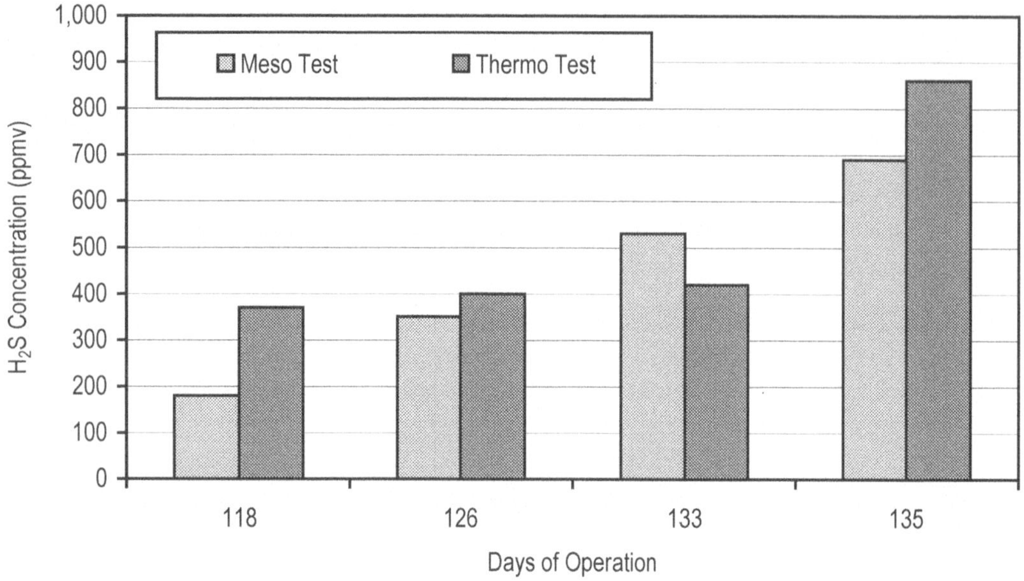

Figure 4-27. Digester Gas Hydrogen Sulfide Concentrations.

The individual VFA analysis on the mesophilic and thermophilic digester sludge samples showed the presence of acetic and propionic acid. Higher molecular weight VFAs (butyric, isobutyric, valeric, and isovaleric) were not detected. In general, the thermophilic digested sludge samples contained higher concentrations of acetic acid (Figure 4-28). Propionic acid was detected in more thermophilic digested sludge samples than mesophilic digested sludge samples; however, the highest propionic acid concentration (280 mg/L) was found in a mesophilic digested sludge sample (Figure 4-29). The Thermo Test propionic acid results were generally lower than those observed in Thermo 49 of Phase 1, Process 2 (Figure 2-27). The lower propionic acid levels in Thermo test could be due to longer HRT and better overall process stability. The individual VFA analytical results are presented in Appendix B (Table B-3).
In summary, Thermo Test (at 50°C) showed higher concentrations of potentially odorous compounds than the Meso Control. This is consistent with literature review findings that thermophilically-digested sludges are more odorous than mesophilically-digested sludges (Garber, 1982; Ghosh et al., 1995; Rimkus et al., 1982; Volpe et al., 1993).

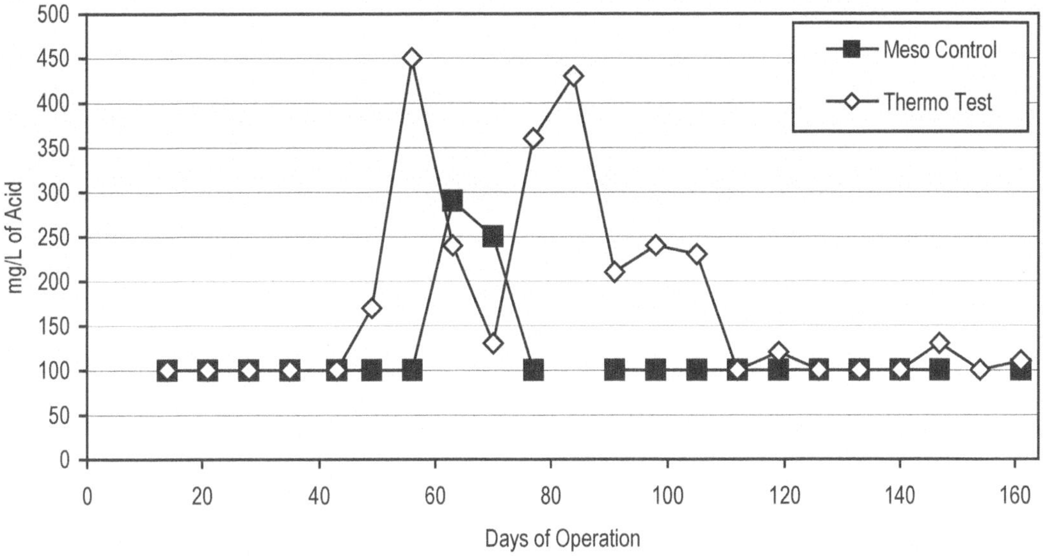

Figure 4-28. Digested Sludge Acetic Acid Concentration.

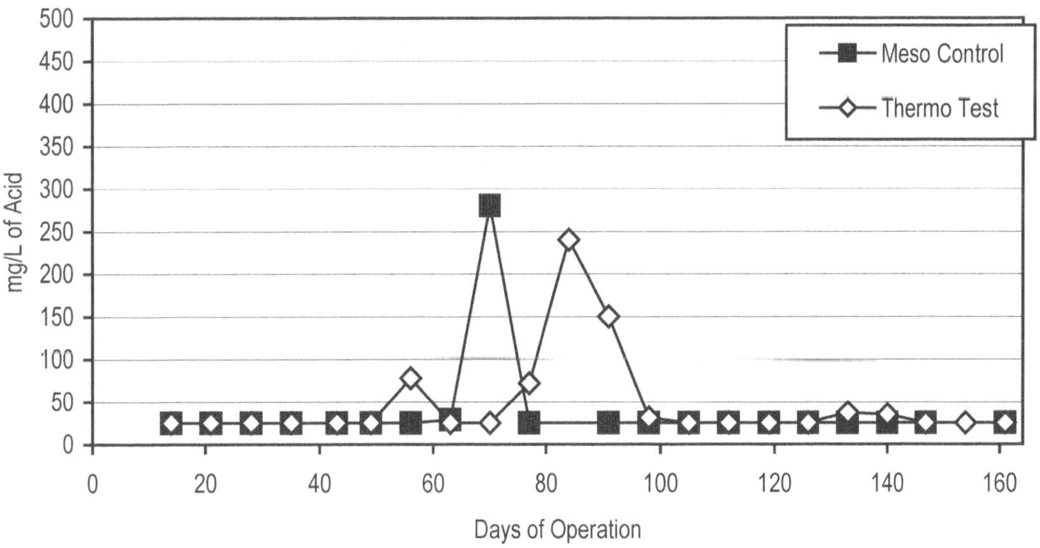

Figure 4-29. Digested Sludge Propionic Acid Concentration.

CHAPTER 5.0

ANALYSIS, SUMMARY, AND CONCLUSIONS

This chapter further analyzes data presented in previous chapters, summarizes previously presented work, and then provides project conclusions and findings.

5.1 Fecal Coliform and Pathogen Destruction

A substantial body of fecal coliform data was collected during each phase of the study, allowing for a comprehensive comparative analysis. An empirical model was developed to make a quantitative comparison between two-stage (Thermo/Meso and Meso/Thermo) and single-stage thermophilic anaerobic systems.

The results reported here are primarily based on data from the following four anaerobic digestion configurations:

1. Mesophilic single-stage (Meso Control);
2. Two-stage, thermophilic acid (2-day MCRT) first stage followed by a mesophilic methane (13- to 15-day MCRT) second stage (Thermo/Meso);
3. Thermophilic single-stage (single-stage Thermo); and
4. Two-stage, mesophilic acid first stage followed by a thermophilic methane second stage (Meso/Thermo).

5.1.1 Fecal Coliform Destruction Model

Using bench-scale and full-scale data collected throughout this study (655 data points), a model that predicts digested sludge (biosolids) fecal coliform density using digester sludge temperature, thermophilic mean cell residence time, staging, and feed sludge fecal coliform density was developed using multiple regression analysis. This model is shown below:

$$Log(FC_f) = 4.19 - 0.0516\ TEMP - 0.0924\ TMCRT - 0.906\ STAGED + 0.211\ Log(FC_i)$$

Where:

FC_f = the digested sludge (biosolids) fecal coliform density (MPN/g TS)

TEMP = the digester sludge temperature (^{o}C)

TMCRT = the digester sludge thermophilic mean cell residence time (days), which includes the MCRT only in thermophilic digesters and excludes the MCRT in mesophilic digesters. [For example a single-stage mesophilic digester with a 20-day MCRT has a TMCRT = 0, and a two-stage Thermo/Meso system with a 2-day MCRT thermophilic first stage and a 13-day MCRT mesophilic second stage has a TMCRT = 2 days.]

STAGED = this value is either 0 or 1, when STAGED = 0 the digester is single stage, when STAGED = 1 the digester is two-stage, with the first stage operated at a 2-day MCRT

FC_i = digester feed sludge fecal coliform density (MPN/g Total Solids)

Table 5-1 shows the results of the regression analysis. The absolute value of each t-statistic for all the coefficients derived in the analysis is much greater than 2 supporting the likelihood that all the coefficients determined in the analysis are true, and the digested sludge fecal coliform density (FC_f) is strongly a function of all the independent variables considered—i.e., TEMP, TMCRT, STAGED, and FC_i (DeLurgio, 1998). The p-values for the coefficients are all less than 1 percent indicating that there is "overwhelming evidence" that the coefficients in the model are valid (Keller and Warrack, 2000), and the probability that any of the coefficients are zero is zero (DeLurgio, 1998). An R^2 of 71.4% means that 71.4% of the variation observed in digested sludge fecal coliform density is explained by the model. The data used to develop the model was tested using methods described in the literature (DeLurgio, 1998; Keller and Warrack, 2000) and found to be suitable for multiple regression analysis. Therefore, the model can be used with reasonable confidence to predict the outcomes of different operating scenarios explored by manipulating the independent variables, alone or in combination.

Table 5-1. Regression Analysis of Phases 1, 2, and 3 Fecal Coliform and Digester Operating Data.
$Log(FC_f) = 4.19 - 0.0516\ TEMP - 0.0924\ TMCRT - 0.906\ STAGED + 0.211\ (FCi)$.

Variable/Constant	Coefficient	t-statistic	p-value
Constant	4.19	14.74	0.000
TEMP	-0.0516	-13.22	0.000
TMCRT	-0.0924	-18.62	0.000
STAGED	-0.906	-12.72	0.000
Log(FCi)	0.211	4.92	0.000

R^2: 71.4%
F-statistic: 405.54
p-value: 0.000

Number of data points = 655

5.1.2 Fecal Coliform Destruction Comparison

Using the model derived from the regression analysis, performance curves can be constructed to predict comparative fecal coliform destruction for each of the thermophilic anaerobic digestion process configurations tested under various conditions. Assuming an initial fecal coliform density of $10^{8.5}$ MPN/g TS, Figure 5-1 predicts the resulting digested sludge (biosolids) fecal coliform density for anaerobic digester systems operated at thermophilic MCRTs from 2–15 days and at 45°C, 50°C, or 55°C. The figure also compares thermophilic systems with and without staging. Since all first stage digesters of the two-staged systems were operated with a 2-day MCRT, the staged systems indicated at a 2-day thermophilic MCRT (x-axis) in Figure 5-1 represent a thermophilic first stage (with a 2-day MCRT) and a mesophilic second stage (with a 13–15 day MCRT), and all other staged systems (thermophilic MCRT = 4–15 days) represent a mesophilic first stage (with a 2-day MCRT) and a thermophilic second stage at the MCRT indicated on the x-axis.

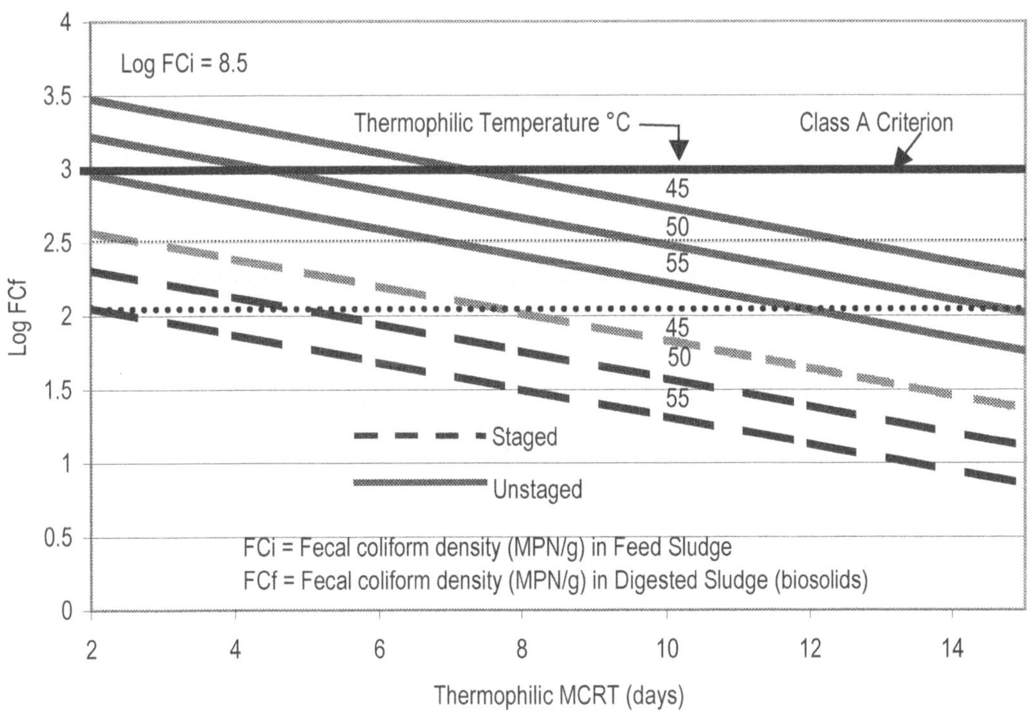

Figure 5-1. Performance Curves Based on Model-Predicted Fecal Coliform Destruction for Thermophilic Anaerobic Digesters Fed Sludge with a Fecal Coliform Density = $10^{8.5}$ MPN/g TS.

These curves show that a two-stage anaerobic digestion system with a 2-day MCRT thermophilic first stage followed by a 13-day MCRT mesophilic second stage does not perform as well as a 15-day MCRT single-stage thermophilic digester. In turn, the 15-day MCRT one-stage thermophilic digester does not perform as well as a two-stage system with a mesophilic first stage (2-day MCRT) followed by a thermophilic second-stage (13-day MCRT), when all thermophilic digesters are operated at the same temperature. This finding is confirmed by the Phase 2 experiments using primary sludge seeded with fecal coliform, poliovirus, and *A. suum,* and thermophilic digesters operating at 50°C (see Figures 5-2 through 5-4).

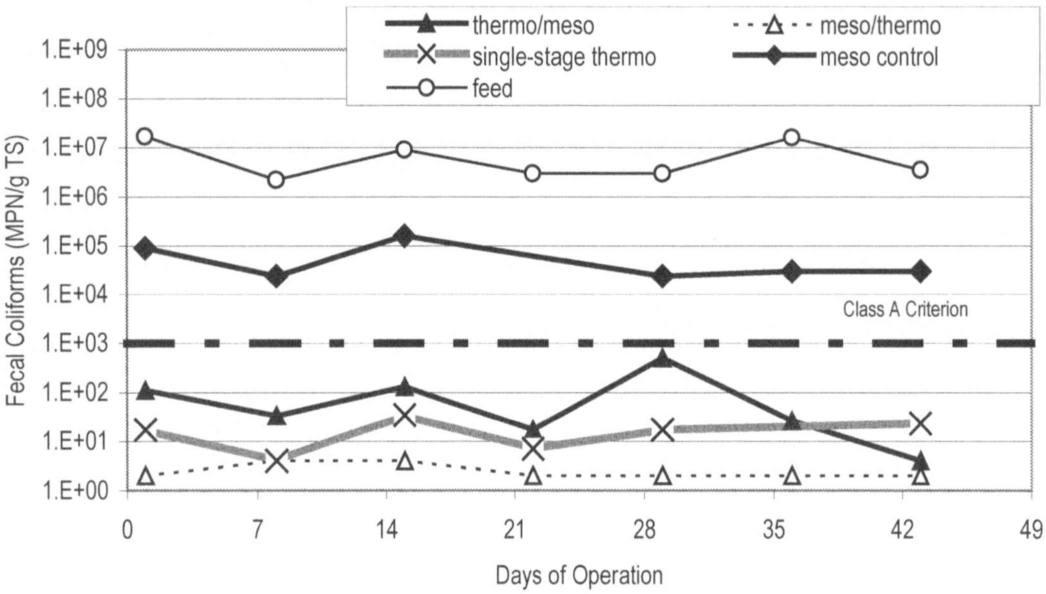

Figure 5-2. Phase 2 Study: Fecal Coliform Results.

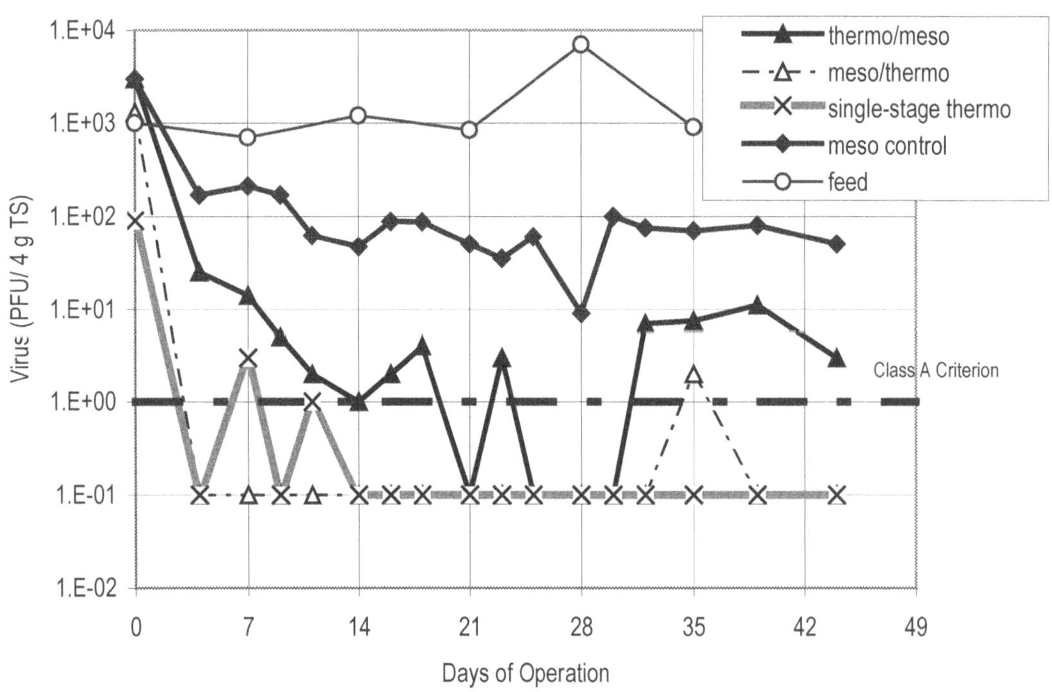

Figure 5-3. Phase 2 Poliovirus Results.

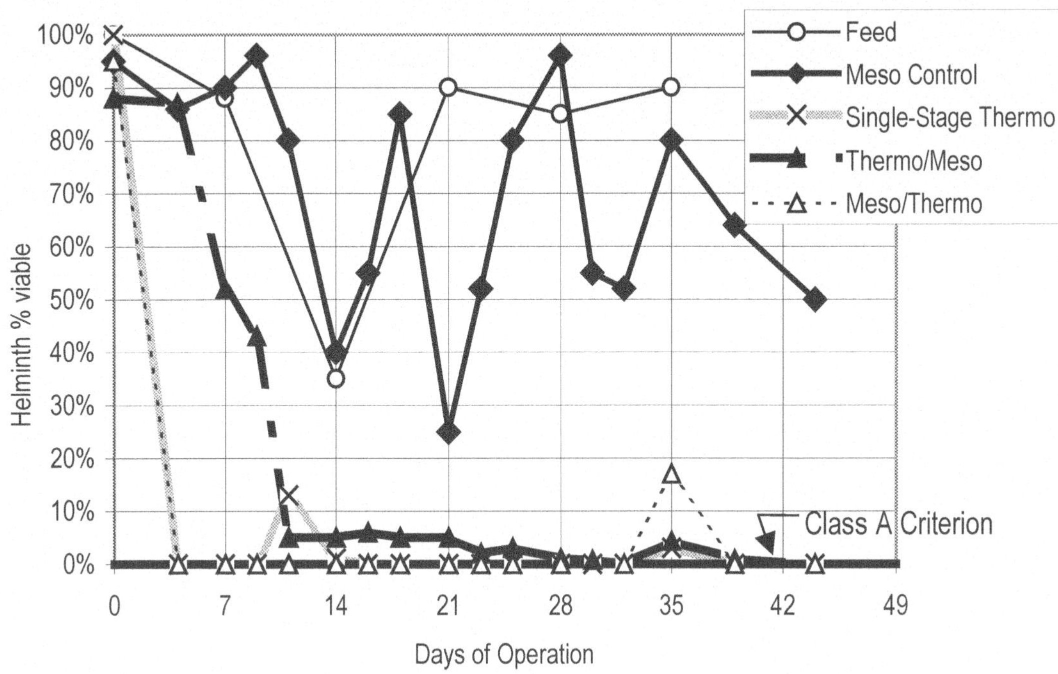

Figure 5-4. Phase 2 Helminth Ova Results.

Figure 5-1 shows that a 2-stage Thermo/Meso digester with a 2-day MCRT thermophilic first stage operated at 55°C and a 13-day MCRT mesophilic second stage (Log $FC_f = 2.06$), reduced fecal coliforms to about the same level as a single-stage thermophilic digester operated at only 50°C with a 15-day MCRT (Log $FC_f = 2.02$). Both easily met the Class A fecal coliform standard (Log $FC_f < 3.00$). For comparison, the Meso/Thermo system with the second-stage thermophilic digester operating at only 45°C and a 12-day thermophilic MCRT would outperform both systems just described (Log $FC_f = 1.65$).

Table 5-2 compares the fecal coliform density predicted by the model with actual results from the Phase 2 study. The average feed sludge fecal coliform density in Phase 2 was $10^{6.7}$ MPN/g TS, with all mesophilic digesters operated at 35°C and all thermophilic digesters operated at 50°C. The first stage digester MCRT was 2 days, and the second stage MCRT was 13 days, for both two-stage digesters. The total MCRT for each of the digester processes tested in Phase 2 was 15 days. The predicted and measured log values for digested sludge fecal coliforms for the two-stage Thermo/Meso system were strikingly close at 1.93 and 1.97 respectively. However, the model predicted higher log values for the two-staged Meso/Thermo and single-stage thermophilic digesters and under predicted this value for the mesophilic single-stage (control) digester than those measured. The predicted value for the mesophilic control digester did, however, fall within the range of actual measured values. Table 5-2 (both predicted and measured values) further verifies the superior performance of the single-stage Thermo over the Thermo/Meso two-stage digester and the Meso/Thermo two-stage over the one-stage Thermo digester.

Table 5-2. Digested Sludge (Biosolids) Fecal Coliform (FC) Densities: Model Predicted vs. Measured in Phase 2. $Log(FC_f) = 4.19 - 0.0516\ TEMP - 0.0924\ TMCRT - 0.906\ STAGED + 0.211\ Log(FCi)$ $Log(FCi) = 6.7$.

Model Inputs and Results					Phase 2 Results		
Digester Configuration	TEMP	TMCRT	STAGED	Log FC Predicted	Log FC Mean	Log FC Range	No. of Samples
Mesophilic Control	35	0	0	3.79	4.35	3.40-5.20	14
Thermo/Meso	50	2	1	1.93	1.97	0.30-2.70	16
Single-Stage Thermo	50	15	0	1.63	1.06	0.30-1.53	13
Meso/Thermo	50	13	1	0.91	0.36	0.30-0.60	14

The Meso/Thermo digester's better performance compared to the single-stage thermophilic digester, even though the Meso/Thermo digester's thermophilic MCRT was less than that of the single-stage thermophilic digester, supports the hypothesis that staging and separating acid and methane phases improves pathogen destruction in anaerobic digestion. This study also showed that a two-stage mesophilic digester (mesophilic first stage followed by a mesophilic second stage) operated in Phase 1 significantly outperformed the single-stage mesophilic control digester, and usually met Class A fecal coliform requirements but not as reliably as the thermophilic digesters (De León et al., 2000). The model presented here also predicts this.

5.1.3 Hold Times

All thermophilic digesters operated in Phases 1, 2, and 3 were operated in the fill/hold/draw mode, not drawing from the digester until the fill and hold cycles were completed. The hold times ranged from 1–6 hours for Phases 1 and 2 (Table 5-3). Although the single-stage thermophilic digesters had hold times of 3 hours (Phase 2) and 6 hours (Phase 1), and the Meso/Thermo digesters had thermophilic hold times of 2 hours (Phase 1) and 6 hours (Phase 2), the thermophilic hold time for the Thermo/Meso was 1 hour during both Phase 1 and 2. The shorter hold time for the Thermo/Meso digester could be responsible for its poorer performance, but hold time does not explain the consistently better performance of the Meso/Thermo digester over the single-stage thermophilic digester—since during Phase 1 the Meso/Thermo digester's hold time was only 2 hours whereas the single-stage thermophilic digester's hold time was 6 hours.

Table 5-3. Thermophilic Digester Hold Times.

Study Phase	Process	Digester	Total MCRT (days)	Hold Time (hrs)
1	Thermophilic 1st Stage	Thermo Acid	2	1
	Single-Stage Thermophilic		15	6
	Meso/Thermo	Meso Acid	2	1
		Thermo methane	15	2
2	Thermo/Meso	Thermo Acid	2	1
		Meso methane	13	6
	Single-Stage Thermophilic		15	3
	Meso/Thermo	Meso Acid	2	1
		Thermo methane	13	6

Phase 3 investigated the single-stage thermophilic anaerobic digestion process in 1.5 million gallon (MG) East Bay Municipal Utility District (EBMUD) digesters. The full-scale digesters were fed intermittently over 20 or more hours without withdrawing sludge from the digester. A 4–6-hour hold period (during which no sludge entered or exited the digester) followed the feed period, and only after this hold period was completed was sludge withdrawn from the digesters. This 24–28 hour feed/hold/draw cycle is repeated every 24–28 hours to treat all the sludge fed to the digester, and is a distinguishing feature of the EBMUD single-stage thermophilic anaerobic digestion process.

Phase 2 demonstrated at bench-scale that the single-stage thermophilic anaerobic digestion process meets Class A pathogen standards by showing sufficient log removals of fecal coliform, enteric virus, and helminth ova by the process (see Figures 5-2 through 5-4). Because helminth ova are either very low or completely absent from EBMUD's digester feed sludge, *A. suum* was added to sludges fed to the bench-scale digesters during Phase 2. Since seeding EBMUD's full-scale digesters with helminth ova is not practical, Sentinel Chambers containing helminth ova were cable hung at approximately 5, 10, and 15 ft below the liquid surface of the digester. The digester was operated at 45°C during the first test using Sentinel Chambers. Two chambers were taken at each depth every 2 hours up to 12 hours. Helminth ova only began to die off after 12 hours at this temperature.

Because of this poor performance, the digester's temperature was increased to 50°C, and approximately 2 months later, the test was rerun. This time chambers were removed at 4, 6, 8, 12, and 24 hours. Four additional chambers were taken at 4, 6, 8, and 24 hours and sent to a different laboratory for analysis. Although the controls showed 95% to 99% survivals, all of the chambers placed in the 50°C, 1.5 MG-digester had 0% survival, except one taken at 12 hours, which had 0.85% survival. This is the only sample out 34 total samples taken that showed any ova viability, and all of the 22 samples taken hours before this sample had 0% viability. It is suspected that this particular sample was contaminated either during sampling or analysis.

The Phase 3 helminth results demonstrated that 4-h hold times would be sufficient in a 50°C thermophilic anaerobic digester to prevent the release of viable helminth ova, even if severe short-circuiting occurred. The Phase 2 results, however, show that a 3-h hold time is sufficient for a 50°C thermophilic anaerobic single-stage digester to meet Class A requirements for fecal coliform, enteric virus, and helminth ova (Figures 5-2 through 5-4), and Phase 1 results show that a Meso/Thermo digester can meet Class A fecal coliform standards with a 2-hour thermophilic hold time (De León et al., 2000). Even the Thermo/Meso digester with only a 1-hour hold time met Class A for fecal coliforms, and mostly met Class A for helminth ova—but only infrequently met Class A for enteric virus.

5.2 Effects of Thermophilic Operation on Process Parameters

5.2.1 Process Performance and Stability

The main points regarding process performance and stability from the Phase 1 Screening Evaluation include the following:

- Acid-phase digesters showed higher pH values and alkalinities at higher temperatures.
- The single-stage digesters showed higher VAs and propionic acid concentrations at higher temperatures (>53°C).
- The methane-phase digesters showed some relationship between temperature and TVA, total alkalinity (TALK), and pH. The pH and TALK tended to decrease with higher temperature, whereas TVAs, including higher molecular weight volatile fatty acids, increased. This may be symptomatic of process instability at the higher temperature. The higher-temperature thermophilic single-stage and methane-phase digesters also showed lower VS reduction.

The screening evaluation allowed comparisons to be made at different digester operating temperatures with a specific process configuration, as well as, to some extent, for different process configurations at the same operating temperature. The single-stage and methane-phase digesters appeared to behave differently at higher temperatures than the acid-phase digesters.

It is difficult to make process comparisons among the digesters tested in Phase 2 of the study because the single-stage and methane-phase digesters were not stable for portions of this relatively brief test (45 days), exhibiting high TVA concentrations and low VS reduction. The digester scale (2 L) may have played some role in the observed instability as it is more difficult to control smaller sludge feed and withdrawal volumes. Also, the emphasis of this phase was on pathogen analysis, and process samples were not collected as frequently as during Phase 1.

Phase 3's primary objective was to confirm pathogen destruction results for a selected thermophilic process at full scale and, to a lesser degree, process performance and stability observed at bench-scale. The single-stage Thermo Test digester was operated at two target operating temperatures: 45°C and 50°C (Periods 1 and 2 respectively). As the Thermo Test temperature increased, the average TVA concentration increased from 200–300 mg/L as acetic acid. This is consistent with observations of the single-stage bench-scale digesters tested in Phase 1. Although on an absolute basis, the TVA levels in the bench-scale Thermo 49 digester were higher than in the full-scale Thermo Test digester during period 2 (average concentrations of 500 mg/L and 300 mg/L, respectively), the TVA/TALK ratios were approximately equivalent (0.06 to 0.07). In general, as the operating temperature of a single-stage thermophilic anaerobic digester increases, higher TVA concentrations and TVA/TALK ratios are expected.

Phase 1 results showed an increase in digester VS reduction when digester temperatures increased from mesophilic (35°C) to 48°C, but VS reduction dropped, even below that of the mesophilic digester, as thermophilic temperatures rose to 52°C and 62°C. This pattern was consistent for both the thermophilic 15-day MCRT single-stage and Meso/Thermo digester configurations, but not for the 2-day MCRT thermophilic digesters (Figure 5-5). This suggests that although increasing digester temperature from mesophilic to the lower end of thermophilic could increase volatile solids reduction, further increasing digester temperatures could reduce TVA utilization to levels lower than associated with mesophilic temperatures. Because anaerobic digestion performance is often quantified by VS reduction (Grady et al., 1999; Metcalf and Eddy, 2003), this also suggests that the anaerobic digestion process may be impaired as digester operating temperatures rise above approximately 53°C. It appears that the temperature at which anaerobic digestion is impacted may be higher for digesters operating at lower MCRTs (Figure 5-5). Other factors, such as feed sludge character, may also influence the temperature at which a process might be affected.

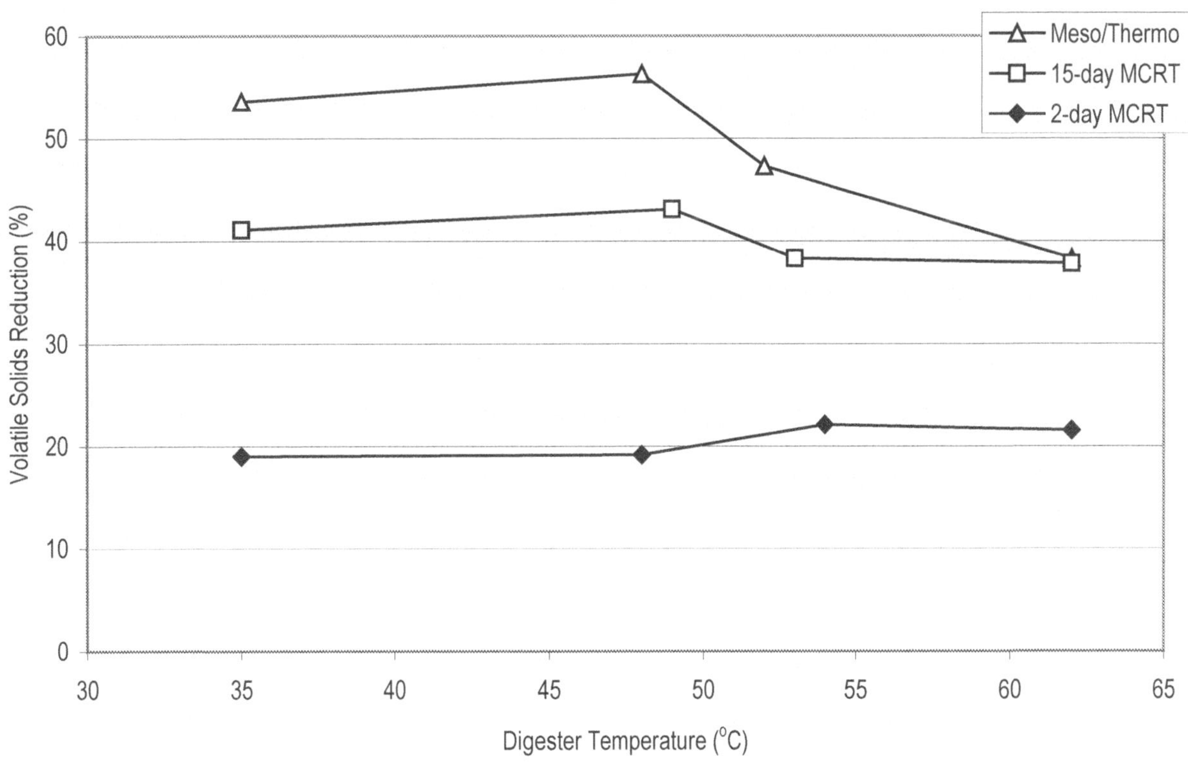

Figure 5-5. Percent Volatile Solids Reduction vs. Digester Temperature.

Schafer et al. (2003) collected VS reduction data from a variety of staged thermophilic anaerobic digester systems. They showed that VS reduction sometimes increased significantly and at other times practically not at all when anaerobic digesters were converted from single-stage mesophilic to staged thermophilic digestion systems. Their findings are strongly supported by the literature. The thermophilic anaerobic digestion literature, however, focuses on thermophilic temperatures between 49°C and 58°C, and mostly at 55°C. Few studies have explored digester VS reduction above 60°C. In contrast, Huyard et al. (1998) showed that VS reduction dropped when the thermophilic temperature in a two-stage Thermo/Meso digester increased from 55° to 60°C, and Garber (1982) showed that VS reduction dropped slightly when digester temperature rose from 36° to 49°C in a single-stage digester. How much or whether VS reduction increases or decreases with increased digester temperature may be a function of feed sludge characteristics.

Volatile solids reduction may also vary according to process configuration (Figure 5-5). The Thermo first stage was not followed by a mesophilic second stage in Phase 1, and the much lower MCRT (2 days compared to 15 days for the other configurations) explains the much lower VS reduction for the Thermo first stage. Further, although Phase 1 digesters that were operated in the same process configuration (i.e., Meso/Thermo, single-stage thermophilic, or first stage thermophilic) were operated side-by-side and received the same feed sludge, digesters with different process configurations were operated at different periods during the year and may have been fed sludges with different characteristics. Feed sludges were made up from a predetermined mixture of EBMUD primary and thickened waste activated sludges, and the characteristics of these sludges are known to vary significantly throughout the year. Therefore, the difference between the Meso/Thermo digester and the single-stage thermophilic digester could be more a function of feed sludge characteristics rather than process configuration.

5.2.2 Determining the Effects of Temperature, pH, Total Volatile Acids, and Total Alkalinity on Digested Sludge Fecal Coliform Density Using Regression Analysis

U.S. EPA 503 regulations consider only time and temperature when evaluating the potential success for thermal disinfecting processes. Most thermophilic anaerobic digesters do not meet U.S. EPA time/temperature requirements but can meet U.S. EPA Class A pathogen requirements according to the 40 CFR 503.32(a)(5) Alternative 3 criteria for reducing pathogenic organisms. Therefore, it has been suggested that other factors found in municipal wastewater treatment plant sludges might enhance disinfection in thermophilic anaerobic digestion and provide for this higher observed level of pathogen destruction.

Digester TVA and ammonia levels have been suggested as lethal factors that may contribute to pathogen disinfection in anaerobic digesters. A regression analysis was performed to investigate the influence of pH, TVA, TALK, temperature (TEMP), and initial fecal coliform density (log FC_i) on digested sludge (biosolids) fecal coliform density using Phase 1 single-stage digester study results. Only single-stage results were used to simplify the analysis.

Table 5-4 presents the results of this multiple regression analysis. T-statistics greater than 2.0 or less than –2.0 (i.e., t-statistics> |2.0|) are considered significant. P-values = 0.000 suggest a very strong statistical correlation between the particular independent variable and the dependent variable. T-statistics and p-values demonstrate that digested sludge fecal coliform densities are strongly correlated to temperature, log FC_i, and pH but are weakly correlated, if at all, to digested sludge TVA or TALK concentrations. The R^2 of 69.9% shows that about 70% of the variation in log FC_i was a result of variation in the independent variables considered.

Table 5-4. Multiple Regression Analysis for Disinfection: Pass 1.
$Log(FC_f) = 12.7 – 0.0806\ TEMP + 0.474\ Log(FC_i) – 1.03\ pH – 0.000058TVA – 0.000157TALK$.

Variable/Constant	Coefficient	t-statistic	p-value
Constant	12.7	6.65	0.000
TEMP	-0.0806	-12.48	0.000
$Log(FC_i)$	0.474	6.91	0.000
pH	-1.03	-4.22	0.000
TVA	-0.000058	-1.01	0.314
TALK	-0.000157	-1.64	0.102

R^2= 69.9%
F-statistic= 96.16

Table 5-5 presents results from the same analysis as Table 5-4 but without VA or TALK. Under this scenario, the t-statistics for temp, pH, and log FC_i improve, and the F-statistic increases from 96.2 to 158.4. These shifts indicate that the overall relationship without inclusion of VA and TALK is stronger. The R^2 changes little, dropping from 69.9% to 69.5%.

Table 5-5. Multiple Regression Analysis for Disinfection: Pass 2.
$Log(FC_f) = 12.0 - 0.0831TEMP + 0.483\ Log(FC_i) – 1.07\ pH$.

Variable/Constant	Coefficient	t-statistic	p-value
Constant	12.0	6.58	0.000
TEMP	-0.0831	-18.53	0.000
$Log(FC_i)$	0.483	7.04	0.000
pH	-1.07	-4.81	0.000

R^2 = 69.5%
F-statistic= 158.44

Table 5-6 includes only temperature and pH as independent variables. T-statistics continue to improve and the F-statistic increases further, but R^2 falls to 62.4%. These results indicate that both temperature and pH strongly influence the variation in log FC_f.

Table 5-6. Multiple Regression Analysis for Disinfection: Pass 3.
$Log(FC_f) = 15.4 - 0.0822TEMP - 1.21\ pH$.

Variable/Constant	Coefficient	t-statistic	p-value
Constant	15.4	8.01	0.000
TEMP	-0.0822	-16.59	0.000
pH	-1.21	-4.96	0.000

R^2= 62.4%
F-statistic= 178.84

Table 5-7 includes only temperature and TVA. The t-statistic for temp is still high; however, T-statistics for TVA is very low with a very high p-value. This suggests that there is a strong relationship between temperature and log FC_f but no relationship between TVA and log FC_f. Table 5-8 shows the results of regressing TALK with Log FC_f. The very low t-statistic and F-statistic and the high p-values suggest no relationship between TALK and log FC_f.

Table 5-7. Multiple Regression Analysis for Disinfection: Pass 4.
$Log(FC_f) = 6.10 - 0.0910TEMP + 0.000046TVA$.

Variable/Constant	Coefficient	t-statistic	p-value
Constant	6.10	20.09	0.000
TEMP	-0.0910	-13.16	0.000
TVA	0.000046	0.76	0.448

R^2= 58.4%
F-statistic= 148.73

Table 5-8. Multiple Regression Analysis for Disinfection: Pass 5.
$\text{Log}(FC_f) = 0.98 - 0.000101\text{TALK}$.

Variable/Constant	Coefficient	t-statistic	p-value
Constant	0.98	0.95	0.343
TALK	0.000101	0.66	0.513

$R^2 = 0.2\%$
F-statistic = 0.43

Results presented in Tables 5-4 through 5-8 indicate that digested sludge fecal coliform densities are lower when pH and temperature are higher but are not affected by TVA or TALK concentrations under the conditions tested. Digester sludge TVA and TALK are typically used as indicators of digester process stability. If TVA increases and TALK decreases significantly, the digester process could be unstable. These data suggest log FC_f is not affected by digester instability, so that disinfection in a thermophilic anaerobic digester may be unaffected during periods of process upset.

These regression results also indicate that log FC_f is lower when digester sludge pH is higher. Other factors that may be lethal to fecal coliforms and pathogens in anaerobic digesters may be enhanced or shut off with increasing and decreasing pH respectively. TVAs and ammonia have been suggested as lethal factors in anaerobic digester sludges, and both TVA's and ammonia become more or less toxic as the pH rises or drops. Based on our regression analyses, TVAs are not lethal (i.e., TVAs do not significantly affect log FC_f, per t-statistic and p-values) in the pH range tested, but ammonia or some other factor may be. Ammonia becomes more lethal at higher pH because more un-ionized ammonia is present. It is the un-ionized ammonia that is toxic. Unfortunately, ammonia was not tested for during the Phase 1 study where the data analyzed was taken.

5.2.3 Determining the Effects of pH, Total Volatile Acids, and Total Alkalinity on Digester Process Stability Using Regression Analysis

Anaerobic digester process stability is often evaluated based on digester sludge TALK, TVA concentration, and pH. To evaluate the effects of operating a digester at thermophilic temperatures compared to mesophilic temperatures, the effects of temperature on these parameters were investigated using regression analysis on single-stage Phase 1 data.

Table 5-9 shows regression results of TVA as a function of digester temperature (TEMP). These results show that TVA concentration is strongly correlated to temperature, with very favorable t- and F-statistics and p-values = 0.000. When digester temperature increases, TVAs can also be expected to increase. This supports data presented in Figures 2-28, 2-45, and 4-14 that show higher acetic, propionic, and other volatile acid levels at higher thermophilic temperatures. This regression analysis included both Meso and Thermo data. If the Meso data is removed from the analysis, the results are essentially the same (Table 5-10). This finding suggests that lower thermophilic temperatures will reduce TVA and possibly odors in digested sludge.

Table 5-9. Multiple Regression Analysis for Process Stability: Pass 1. Thermophilic and Mesophilic Data. TVA = -2688 + 77.3TEMP.

Variable/Constant	Coefficient	t-statistic	p-value
Constant	-2688	-9.33	0.000
TEMP	77.3	13.52	0.000

R^2 = 46.2%
F-statistic = 182.78

Table 5-10. Multiple Regression Analysis for Process Stability: Pass 2. Thermophilic Data Only. TVA = -6365 + 143TEMP.

Variable/Constant	Coefficient	t-statistic	p-value
Constant	-6365	-9.09	0.000
TEMP	143	11.21	0.000

R^2 = 44.8%
F-statistic = 125.63

Table 5-11 shows regression results for TALK as a function of TEMP. Although the R^2 was only 4.9%, the t- and p-values both suggest a significant relationship. Eliminating the mesophilic data (Table 5-12), the relationship appears stronger with the R^2 up to 10.7%, the t-statistic for TEMP up to –4.30, and p-value = 0.000. The analysis suggests that digester sludge alkalinity will decrease when digester sludge temperature is increased.

Table 5-11. Multiple Regression Analysis for Process Stability: Pass 3. Thermophilic and Mesophilic Data. TALK = 7267 – 11.3TEMP.

Variable/Constant	Coefficient	t-statistic	p-value
Constant	7267	42.21	0.000
TEMP	11.3	-3.31	0.001

R^2 = 4.9%
F-statistic = 10.97

Table 5-12. Multiple Regression Analysis for Process Stability: Pass 4. Thermophilic Data Only. TALK = 8386 – 31.2TEMP.

Variable/Constant	Coefficient	t-statistic	p-value
Constant	8386	21.02	0.000
TEMP	-31.2	-4.30	0.000

R^2= 10.7%
F-statistic= 18.51

Regression results of pH and TEMP (Table 5-13) show a similar correlation between pH and TEMP as with TALK and TEMP, but if the mesophilic data are removed from the analysis (Table 5-14), the very low t-statistic (0.41), very high p-value (0.680), and $R^2 = 0.1\%$ strongly suggest that there is no relationship between thermophilic temperature and pH.

Table 5-13. Multiple Regression Analysis for Process Stability: Pass 5. Thermophilic and Mesophilic Data. pH = 7.82 + 0.00426TEMP.

Variable/Constant	Coefficient	t-statistic	p-value
Constant	7.82	113.82	0.000
TEMP	0.00426	3.13	0.000

R^2= 4.4%
F-statistic= 9.77

Table 5-14. Multiple Regression Analysis for Process Stability: Pass 6. Thermophilic Data Only. pH = 7.99 + 0.00120TEMP.

Variable/Constant	Coefficient	t-statistic	p-value
Constant	7.99	50.25	0.000
TEMP	0.00120	0.41	0.680

R^2= 0.1%
F-statistic= 0.17

5.3 Dewatering

In a full-scale field test using solid bowl centrifuges, Thermo Test (50°C) sludge dewatered more easily than the Meso Control (35°C) sludge, as shown by the higher cake solids (average values of 24.4% and 20.9%, respectively) and greater solids capture efficiency (average values of 87.7% and 74.2%, respectively). Thermo Test also produced cake with fecal coliform concentrations that are almost 2 logs less than Meso Control and nearly always less than the Class A fecal coliform criterion. However, these dewatering results were contradicted by the CST test, a surrogate indicator of dewaterability. Although CST results indicate that Meso Control digested sludge should dewater more readily than Thermo Test, actual field results indicate the contrary. The CST test may not be a suitable indicator of dewaterability when comparing mesophilic and thermophilic sludges.

The dewatering test conditions were not typical of standard EBMUD operations and may not be representative of other treatment plants. Ordinarily, primary digester sludges at EBMUD are combined in a secondary digester, unmixed and unheated. Under typical operating conditions, mesophilic and thermophilic sludges are combined and allowed to cool to ambient temperatures prior to dewatering. Cooler sludges are more viscous, and it is expected that cake solids concentrations for both thermophilic and mesophilic digested sludges might be lower than those shown during this test had they been stored in a secondary digester.

5.4 Odors

Higher volatile acid levels, especially for propionic acid, in thermophilic sludges have been implicated as the cause of thermophilic sludges being more odorous than mesophilic sludges (Ghosh et al., 1995; Garber, 1982, 1977; and Rimkus et al., 1982). Figure 5-6 shows that digesters operated at higher thermophilic temperatures (i.e., 53°C and 62°C) produce significantly more volatile acids, especially propionic acid, than a digester operated at mesophilic temperature (35°C). However, for digesters operated at 35°C and 49°C, individual volatile acids are similar. This suggests that more odors are generated at the higher thermophilic temperatures, but lower thermophilic temperatures are no more odorous than mesophilic temperatures. Samples were taken from stable single-stage bench-scale digesters after 3 MCRTs of operating time at the temperature indicated. With odor often being a major complaint of thermophilic digesters, operating digesters at the lower end of the thermophilic temperature range would be beneficial.

In full-scale studies, volatile organic compound (VOC) levels were higher in the Thermo Test (50°C) digester than in the Meso Control (35°C). Digested sludge acetic acid concentrations were also higher in Thermo Test than in Meso Control. Results for propionic acid were mixed with a greater number of measurable concentrations from Thermo Test, but the highest single concentration was observed in Meso Control. Higher hydrogen sulfide concentrations were measured in Thermo Test digester gas in three out of four tests, and in three out of four tests no other reduced sulfur compounds were detected. Bench-scale tests suggest that the differences between Thermo Test and Meso Control may have been greater if Thermo Test were operated at a higher thermophilic temperature.

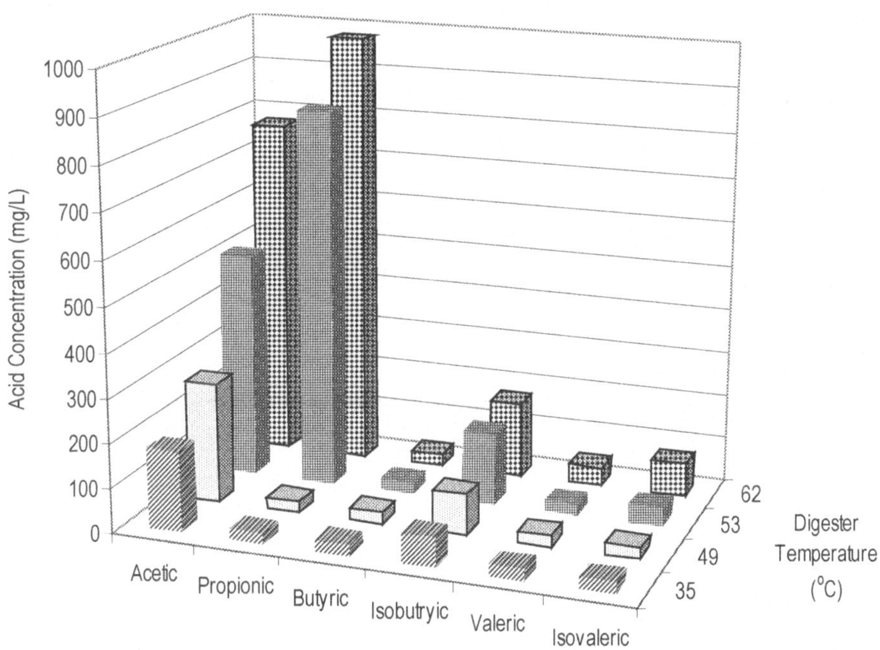

Figure 5-6. Volatile Acid Species Distribution vs. Digester Temperature for Single-Stage Processes, 3 MCRTs After Start-up.

5.5 Conclusions

5.5.1 Principal Conclusions

The principal conclusions from this research include:

- Staging anaerobic digesters improves fecal coliform destruction; however, single-stage thermophilic digesters operating in the fill/hold/draw sequence can outperform two-stage digesters, by further reducing fecal coliforms, when single-stage digesters are operated at higher temperatures or longer thermophilic MCRTs or both. This relationship is quantified by a model presented in this report.

- Anaerobic digesters operated at lower thermophilic temperatures (45–50°C) may generate less odor than digesters operated at higher thermophilic temperatures (53–62°C) because less volatile acids (especially propionic acid) are formed at lower thermophilic temperatures.
- Anaerobic digesters operated at lower thermophilic temperatures (45–50°C) may provide for higher VSR than digesters operated at the higher thermophilic temperatures (53–62°C).
- Anaerobic digesters operated at lower thermophilic temperatures can reliably meet U.S. EPA Alternatives 3 and 4 Class A criteria if an adequate thermophilic MCRT is maintained. Because of its other potential benefits (e.g., less odor, higher VS reduction, and less energy consumed) operating anaerobic digesters in the lower thermophilic temperature range (45–50°C) may be preferable to operating at higher thermophilic temperatures (53–62°C).
- Depending on the thermophilic MCRT and temperature used, this study suggests that the three anaerobic digester configurations evaluated here (i.e., single-staged thermophilic or two-staged—Meso/Thermo or Thermo/Meso) have the potential to achieve U.S. EPA Class A pathogen requirements. This may provide wastewater treatment agencies with the flexibility to select a thermophilic system more adaptable to their existing facilities, rather than being forced to accept a particular thermophilic process.

Additional conclusions are provided by chapter.

5.5.2 Phase 1 Screening Evaluation Conclusions (Chapter 2.0)

- At a 2-day MCRT, when digester thermophilic temperature rises:
    - TVA decreases
    - TALK increases
    - VSR increases
- At a 15-day MCRT, when digester thermophilic temperature rises, in general:
    - TVA increases
    - TALK decreases
    - VSR decreases
    - Specific VFAs (such as propionic acid, which is linked to odor generation) increase
- Two-stage Meso/Meso (mesophilic acid first-stage followed by a mesophilic methane second-stage) meets Class A requirements for fecal coliform (Figure 2-47), but could not meet Class A helminth ova requirements when ova-filled sentinel chambers were held in the process for 17 days (Figure 2-53).

5.5.3 Phase 2 Pathogen Destruction Conclusions (Chapter 3.0)

- Ammonia levels in digester sludge are significantly higher than that in feed sludge. This is probably due to deamination of organic nitrogen from the feed sludge in the digester.
- Digester sludge ammonia levels increase when the digester MCRT is increased from 2 to 15 days, and both systems are fed the same feed sludge.

- Digester sludge ammonia concentration increases with increased digester temperature (from 35° to 50°C) at a 2-day MCRT. The digester sludge ammonia concentration appears to be independent of digester temperature at the longer MCRTs (i.e., 15 days and greater).
- Regressing Phase 2 data showed that the digested sludge (biosolids) fecal coliform density is significantly influenced by digester temperature, feed sludge fecal coliform density, thermophilic MCRT, and whether the digester is two- or single-staged.
- Regressing Phase 2 data showed that biosolids fecal coliform density is not significantly influenced by digester sludge volatile acid or alkalinity levels. Since anaerobic digestion process stability is usually determined by volatile acid to alkalinity ratio (< 0.10 is a stable process), this suggests that thermophilic anaerobic disinfection may not depend on process stability.

5.5.4 Phase 3 Full-Scale Pilot Test Conclusions (Chapter 4.0)

- Thermophilic sludge dewatered better, producing higher cake solids (23% to 27%) and solids capture efficiency (84% to 94%), than mesophilic sludge (cake solids: 20% to 22%, solids capture efficiency: 66% to 88%) in full-scale tests at the EBMUD MWWTP. CST tests predicted the opposite, suggesting that the CST test may not be a good predictor of thermophilically digested sludge dewatering performance.
- VOCs were significantly higher in the full-scale thermophilic digester (4,000–6,000 ppbv at 50°C) compared to the full-scale mesophilic control digester (1,000–4,000 ppbv at 35°C). Sulfide levels in the thermophilic digester gas (512 ppm average of four samples) were somewhat higher than that the mesophilic digester gas (438 average of four samples).
- Sentinel Chamber tests showed that all helminth ova were killed after being held for 4 hours in the full-scale digester operated at 50°C, but there was very little ova kill at 45°C after 12 hours.
- The full-scale thermophilic digester consistently met Class A fecal coliform requirements when operated at 50°C.
- The full-scale thermophilic digester consistently met U.S. EPA VAR requirement of 38% or more VSR.
- The full-scale thermophilic digester occasionally exceeded the TVA/TALK ratio of the mesophilic control, but did not typically exceed 0.10 (generally accepted as stable operation for mesophilic digesters).

5.5.5 Data Analysis Conclusions (Chapter 5.0)

- Regressing data from single-stage digesters all operated at a 15-day MCRT (from Phase 1 data) showed that temperature and pH had the most influence on biosolids fecal coliform densities. This suggests that another disinfecting agent that becomes more lethal as pH rises may play a role in reducing fecal coliforms. This disinfecting agent may be ammonia, since the unionized ammonia fraction of measured ammonia-nitrogen is toxic and increases with increasing pH.
- Regressing the single-stage Phase 1 data showed a strong relationship between increasing digester TVA and increasing digester thermophilic temperatures. This is consistent with specific VFA concentrations measured in Phase 1, where propionic and other acid levels increased as thermophilic temperatures rose to 53–62°C.
- Regression analysis suggests a weak relationship between digester sludge alkalinity and temperature. Alkalinity decreases with increasing thermophilic temperature for single-stage and methane phase digesters.
- Regression analysis suggests that there is no relationship between digester sludge pH and thermophilic digester temperature.

REFERENCES

Aitken, M.D. and Mullennix, R.W. (1992) Another Look at Thermophilic Anaerobic Digestion of Wastewater Sludge. *Water Environ. Res.*, **64**, 915.

American Public Health Association. (1998) *Standard Methods for the Examination of Water and Wastewater,* 20th ed. Washington, DC: American Public Health Association, American Water Works Association, and Water Environment Federation.

Berg, G. and Berman, D. (1980) Destruction by Anaerobic Mesophilic and Thermophilic Digestion of Viruses and Indicator Bacteria Indigenous to Domestic Sludges. *Appl. Environ. Microbiol.*, **39**, 361.

De León, C., Hake, J., Gabb, D., Danielson, R., Ghosh, S., and Williams, D. (2000) Comparison of Thermophilic Anaerobic Digestion Processes for Meeting Biosolids Class A Pathogen Standards I. *Proceedings of the 73rd Annual Water Environment Federation Technical Exposition and Conference* [CD-ROM]; Anaheim, CA.

DeLurgio, S.A. (1998) *Forecasting Principles and Applications*; Irwin/McGraw-Hill: New York.

Drury, D., Baker, C., and Berch, C. (2001) Increased Sludge Dewaterability Through Implementation of Full-Scale, Three-Stage Digestion. California Water Environment Association Bulletin. Spring 2001.

Feachem, R., Bradley, D., Garelick, H., and Mara, D. (1983) Sanitation and Disease Health Aspects of Excrete and Wastewater Management; New York, 375.

Ferran, B., Huyard, A., and Adamik, J. (2002) Two-Phase Anaerobic Digestion of Municipal Sewage Sludge Optimization of the Pathogen Destruction. Proceedings of the 75th Annual Water Environment Federation Technical Exposition and Conference [CD-ROM]; Chicago, IL.

Garber, W.F. (1954) Plant-Scale Studies of Thermophilic Digestion at Los Angeles. *Sewage Ind. Wastes*, **26**, 1202.

Garber, W.F., Ohara, G.T., Colbaugh, J.E., and Raksit, S.K. (1975) Thermophilic Digestion at the Hyperion Treatment Plant. *J. Water Pollut. Control Fed.*, **47**, 950.

Garber, W.F. (1977) Certain Aspects of Anaerobic Digestion of Wastewater Solids in the Thermophilic Range at the Hyperion Treatment Plant. *Prog. Wat. Technol.*, **8**, 401.

Garber, W.F. (1982) Operating Experience with Thermophilic Anaerobic Digestion. *J. Water Pollut. Control Fed.*, **54**, 1170.

Ghosh, S., Buoy, K., Dressel, L., Miller, T., Wilcox, G., and Loos, D. (1995) Pilot- and Full-Scale Two-Phase Anaerobic Digestion of Municipal Sludge. *Water Environ. Res.*, **67**, 206.

Golueke, C.G. (1958) Temperature Effects on Anaerobic Digestion of Raw Sewage Sludge. *Sewage Ind. Wastes*, **30**, 1225.

Grady, L.P.C., Daigger, G.T., and Lim, H.C. (1999) *Biological Wastewater Treatment*; Marcel Dekker, Inc.: New York, 639.

Huyard, A., Ferran, B., Audic, J.M., Dieterien, J., Adamik, J., and Noel, T. (1998) A Challenge for the Two-Phased Anaerobic Digestion: To Produce Class A Biosolids and Meet PFRP Equivalency. Presented at WEFTEC 98, Orlando, FL. Oct. 1998.

Keller, G. and Warrack, B. (2000) *Statistics for Management and Economics*; Belmont, CA: Duxbury Press.

Lee, H., Yuan, H., Netto, H., Liang, J., Fan, S., and Mundine, J. (2001) Anaerobic Thermophilic Digestion Hyperion Treatment Plant Experience. California Water Environment Association Bulletin, Fall 2001, pgs. 28-33.

Lee, K.M., Brunner, C.A., Farrell, J.B., and Eralp, A.E. (1989) Destruction of Enteric Bacteria and Viruses During Two-Phase Digestion. *J. Water Pollut. Control Fed.*, **61**, 1421.

Metcalf & Eddy, Inc. (2003) *Wastewater Engineering—Treatment and Reuse (4th ed.).* New York: McGraw-Hill.

Metcalf & Eddy, Inc. (1991) *Wastewater Engineering—Treatment, Disposal, and Reuse (3rd ed.).* New York: McGraw-Hill, 425.

Pecson, B. and Nelson, K. (2003) The Effects of Exposure Time, Temperature, pH, and Ammonia Concentration on the Inactivation Rate of *Ascaris* Eggs. Proceedings of the 76th Annual Water Environment Federation Technical Exposition and Conference [CD-ROM]; Los Angeles, CA.

Rimkus, R.R., Ryan, J.M., and Cook, E.J. (1982) Full-Scale Thermophilic Digestion at the West-Southwest Sewage Treatment Works, Chicago, Illinois. *J. Water Pollut. Control Fed.*, **54**, 1447.

Schafer, P. and Farrell, J. (2000) Turn Up the Heat. *Water Environment and Technology*, **12** (11).

Schafer, P., Farrell, J., Newman, G., and Vandenburgh, S. (2003) Advanced Anaerobic Digestion Processes, *Water Environment and Technology*, **15** (5).

Stern, G., and Farrell, J.B. (1977) Sludge Disinfection Techniques. In: *Proc. Nat. Conf. on Composting of Municipal Residues and Sludges*. Rockville, MD: Pub. Information Transfer, Inc., 142.

U.S. EPA (1999) *Control of Pathogens and Vector Attraction in Sewage Sludge*. EPA Pub. No. 625/R-92/013. Office of Research and Development, Washington, D.C.

U.S. EPA (1994) *Standards for the Use or Disposal of Sewage Sludge*. 40 CFR Part 503 (revised July 1, 1994).

U.S. EPA (1989) *Control of Pathogens in Municipal Wastewater Sludge*. EPA Pub. No. 625/10-89/006. EPA Center for Environmental Research Information, Cincinnati, OH.

Volpe, G., Rabinowitz, B., Peddie, C., and Krugel, S. (1993) Class A (High Grade) Sludge Process Design for the Greater Vancouver Regional District (GVRD) Annacis Island Wastewater Treatment Plant. In: *Proc. Water Env. Fed. 66th Annual Conf. and Expo. (Sludge Management Vol. 4.)*, Anaheim, CA, 1.

Ward, R.L., and Ashley, C.S. (1978) Methods to Inactivate Enteric Viruses in Wastewater Sludge. Sandia Laboratories (Sand77-1498). Albuquerque, NM, 1.

Yang, Y.C. (1996) "Abiotic Factors in Biosolids Processing that Influence Pathogen Destruction (Nitrous Acid and Ammonia Studies)," MS Thesis, Tulane University School of Public Health and Tropical Medicine, New Orleans, LA.

APPENDIX A

Photographs of Bench-Scale Digesters

Figure A-1. Phase 1 30-L Anaerobic Digesters.

Figure A-2. Phase 1 Feed Sludge Preheating with 35°C Water Bath.

Figure A-3. Phase 2 1.5-L Anaerobic Digester System.

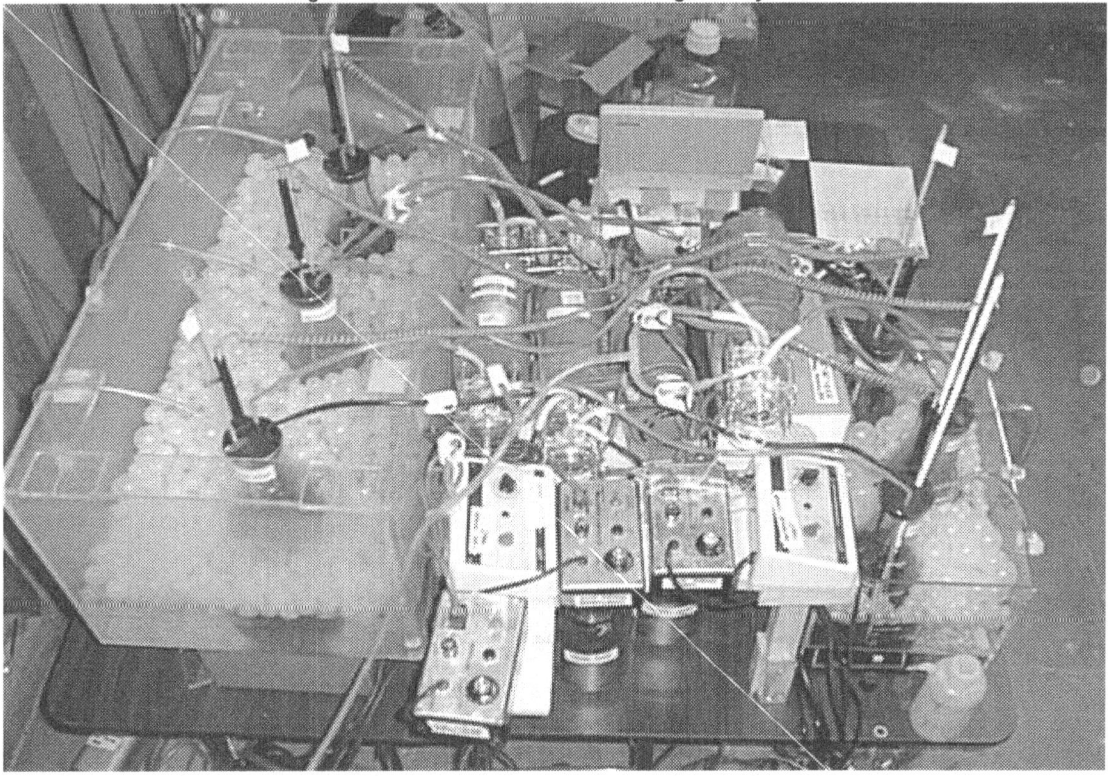

Figure A-4. Overview of Phase 2 Digester System.

APPENDIX B

ODOR DATA
LIST OF COMPOUNDS SCANNED

Table B-1. VOC Results.

Days of Operation	Day 118				Day 126				Day 133				Day 135			
	Meso		Thermo		Meso		Thermo		Meso		Thermo		Meso		Thermo	
Compound	ppb	mg/m^3	ppb	mg/m^3	ppb	mg/m^3	ppb	mg/m^3	ppb	mg/m^3	ppb	mg/m^3	ppb	mg/m^3	ppb	mg/m^3
Freon 12	ND	ND	ND	ND	ND	ND	ND	ND	ND	ND	ND	ND	ND	ND	ND	ND
Freon 114	ND	ND	ND	ND	ND	ND	ND	ND	ND	ND	ND	ND	ND	ND	ND	ND
Chloromethane	ND	ND	ND	ND	ND	ND	ND	ND	ND	ND	ND	ND	ND	ND	ND	ND
Vinyl Chloride	ND	ND	ND	ND	ND	ND	ND	ND	ND	ND	ND	ND	ND	ND	ND	ND
Bromomethane	ND	ND	ND	ND	ND	ND	ND	ND	ND	ND	ND	ND	ND	ND	ND	ND
Chloroethane	ND	ND	ND	ND	ND	ND	ND	ND	ND	ND	ND	ND	ND	ND	ND	ND
Freon 11	ND	ND	ND	ND	ND	ND	ND	ND	ND	ND	ND	ND	ND	ND	ND	ND
1,1-Dichloroethene	2.5	10	ND	ND	ND	ND	ND	ND	3.1	12	ND	ND	ND	ND	ND	ND
Freon 113	ND	ND	ND	ND	ND	ND	ND	ND	ND	ND	ND	ND	ND	ND	ND	ND
Methylene Chloride	ND	ND	ND	ND	ND	ND	ND	ND	ND	ND	7.3	26	ND	ND	ND	ND
1,1-Dichloroethane	ND	ND	ND	ND	ND	ND	ND	ND	ND	ND	ND	ND	ND	ND	ND	ND
cis-1,2-Dichloroethene	8.6	35	27	110	14	57	43	170	21	86	99	400	23	94	30	120
Chloroform	ND	ND	ND	ND	ND	ND	ND	ND	ND	ND	ND	ND	ND	ND	ND	ND
1,1,1-Trichloroethane	ND	ND	ND	ND	ND	ND	ND	ND	ND	ND	ND	ND	ND	ND	ND	ND
Carbon Tetrachloride	ND	ND	ND	ND	ND	ND	ND	ND	ND	ND	ND	ND	ND	ND	ND	ND
Benzene	7.3	24	17	55	ND	ND	19	61	5.4	18	20	64	ND	ND	22	72
1,2-Dichloroethane	ND	ND	ND	ND	ND	ND	ND	ND	ND	ND	ND	ND	ND	ND	ND	ND
Trichloroethene	25	140	39	210	43	240	68	370	43	240	84	460	46	250	60	320
1,2-Dichloropropane	ND	ND	ND	ND	ND	ND	ND	ND	ND	ND	ND	ND	ND	ND	ND	ND
cis-1,3-Dichloropropene	ND	ND	ND	ND	ND	ND	ND	ND	ND	ND	ND	ND	ND	ND	ND	ND
Toluene	340	1300	560	2100	340	1300	430	1700	320	1200	430	1600	350	1300	400	1500
trans-1,3-Dichloropropene	ND	ND	ND	ND	ND	ND	ND	ND	ND	ND	ND	ND	ND	ND	ND	ND
1,1,2-Trichloroethane	ND	ND	ND	ND	ND	ND	ND	ND	ND	ND	ND	ND	ND	ND	ND	ND
Tetrachloroethene	18	120	23	160	65	450	30	200	54	370	52	360	46	320	49	340
1,2-Dibromoethane (EDB)	ND	ND	ND	ND	ND	ND	ND	ND	ND	ND	ND	ND	ND	ND	ND	ND
Chlorobenzene	7.1	33	15	69	ND	ND	14	68	7.4	35	13	63	ND	ND	ND	ND
Ethyl Benzene	40	180	110	500	63	280	130	590	160	690	180	800	150	680	140	620
m,p-Xylene	42	180	110	500	62	270	120	530	64	280	140	620	70	310	120	530
o-Xylene	16	70	40	170	21	93	47	210	24	110	52	230	25	110	41	180
Styrene	2.3	10	10	45	ND	ND	ND	ND	12	50	24	100	ND	ND	21	89
1,1,2,2-Tetrachloroethane	ND	ND	ND	ND	ND	ND	ND	ND	ND	ND	ND	ND	ND	ND	ND	ND
1,3,5-Trimethylbenzene	17	83	41	210	34	170	62	310	30	150	66	330	36	180	56	280
1,2,4-Trimethylbenzene	35	170	94	470	72	360	140	720	77	380	190	960	83	420	140	710
1,3-Dichlorobenzene	ND	ND	ND	ND	ND	ND	ND	ND	ND	ND	ND	ND	ND	ND	ND	ND
1,4-Dichlorobenzene	13	77	57	350	41	250	99	610	35	220	140	840	51	310	99	600
alpha-Chlorotoluene	ND	ND	ND	ND	ND	ND	ND	ND	ND	ND	ND	ND	ND	ND	ND	ND
1,2-Dichlorobenzene	ND	ND	ND	ND	ND	ND	ND	ND	ND	ND	ND	ND	ND	ND	ND	ND
1,2,4-Trichlorobenzene	ND	ND	ND	ND	ND	ND	ND	ND	ND	ND	ND	ND	ND	ND	ND	ND
Hexachlorobutadiene	ND	ND	ND	ND	ND	ND	ND	ND	ND	ND	ND	ND	ND	ND	ND	ND
Propylene	430	750	1100	2000	480	840	1700	2900	ND	ND	ND	ND	290	510	2300	4000

Table B-1. VOC Results, continued.

Days of Operation	Day 118				Day 126				Day 133				Day 135			
	Meso		Thermo		Meso		Thermo		Meso		Thermo		Meso		Thermo	
Compound	ppb	mg/m^3	ppb	mg/m^3	ppb	mg/m^3	ppb	mg/m^3	ppb	mg/m^3	ppb	mg/m^3	ppb	mg/m^3	ppb	mg/m^3
1,3-Butadiene	ND	ND	ND	ND	ND	ND	ND	ND	ND	ND	ND	ND	ND	ND	ND	ND
Acetone	19	45	640	1500	46	110	260	620	13	32	320	770	ND	ND	150	350
Carbon Disulfide	14	43	42	130	ND	ND	120	390	13	42	110	360	ND	ND	ND	ND
2-Propanol	27	67	96	240	ND	ND	50	120	19	48	ND	ND	ND	ND	120	310
trans-1,2-Dichloroethene	ND	ND	ND	ND	ND	ND	ND	ND	ND	ND	ND	ND	ND	ND	ND	ND
Vinyl Acetate	ND	ND	ND	ND	ND	ND	ND	ND	ND	ND	ND	ND	ND	ND	ND	ND
2-Butanone (MEK)	ND	ND	460	1400	ND	ND	290	870	12	37	470	1400	ND	ND	140	430
Hexane	16	56	20	70	ND	ND	ND	ND	18	66	ND	ND	ND	ND	140	510
Tetrahydrofuran	ND	ND	ND	ND	ND	ND	ND	ND	ND	ND	ND	ND	ND	ND	ND	ND
Cyclohexane	ND	ND	ND	ND	ND	ND	ND	ND	9.4	33	ND	ND	ND	ND	ND	ND
1,4-Dioxane	ND	ND	ND	ND	ND	ND	ND	ND	ND	ND	ND	ND	ND	ND	ND	ND
Bromodichloromethane	ND	ND	ND	ND	ND	ND	ND	ND	ND	ND	ND	ND	ND	ND	ND	ND
4-Methyl-2-pentanone	11	48	230	970	ND	ND	240	1000	ND	ND	290	1200	ND	ND	110	440
2-Hexanone	ND	ND	30	120	ND	ND	ND	ND	ND	ND	40	160	ND	ND	ND	ND
Dibromochloromethane	ND	ND	ND	ND	ND	ND	ND	ND	ND	ND	ND	ND	ND	ND	ND	ND
Bromoform	ND	ND	ND	ND	ND	ND	ND	ND	ND	ND	ND	ND	ND	ND	ND	ND
4-Ethyltoluene	42	210	110	530	76	380	140	720	85	430	180	880	87	440	130	640
Ethanol	ND	ND	28	54	ND	ND	ND	ND	ND	ND	1000	2000	2400	4600	1400	2700
Methyl tert-Butyl Ether	12	45	22	81	ND	ND	ND	ND	15	54	28	100	ND	ND	ND	ND
Heptane	48	200	69	290	ND	ND	ND	ND	65	270	67	280	250	1100	510	2100
TOTAL	1,193	3,896	3,990	12,334	1,357	4,800	4,002	12,159	1,105	4,853	4,002	14,003	3,907	10,624	6,178	16,841

ND non-detect

Table B-2. Sulfur Compound Results.

Compound	Sulfur Compound Concentration (ppmv)							
	Day 118		Day 126		Day 133		Day 135	
	Meso	Thermo	Meso	Thermo	Meso	Thermo	Meso	Thermo
Hydrogen Sulfide	180	370	350	400	530	420	690	860
Carbonyl Sulfide	ND	8.1	ND	ND	ND	ND	ND	ND
Methyl Mercaptan	ND	ND	ND	ND	ND	ND	ND	ND
Ethyl Mercaptan	ND	ND	ND	ND	ND	ND	ND	ND
Dimethyl Sulfide	ND	ND	ND	ND	ND	ND	ND	ND
Carbon Disulfide	ND	ND	ND	ND	ND	ND	ND	ND
Isopropyl Mercaptan	ND	ND	ND	ND	ND	ND	ND	ND
tert-Butyl Mercaptan	ND	ND	ND	ND	ND	ND	ND	ND
n-Propyl Mercaptan	ND	ND	ND	ND	ND	ND	ND	ND
Ethyl Methyl Sulfide	ND	ND	ND	ND	ND	ND	ND	ND
Thiophene	ND	ND	ND	ND	ND	ND	ND	ND
Isobutyl Mercaptan	ND	ND	ND	ND	ND	ND	ND	ND
Diethyl Sulfide	ND	ND	ND	ND	ND	ND	ND	ND
Butyl Mercaptan	ND	ND	ND	ND	ND	ND	ND	ND
Dimethyl Disulfide	ND	ND	ND	ND	ND	ND	ND	ND
3-Methylthiophene	ND	ND	ND	ND	ND	ND	ND	ND
Tetrahydrothiophene	ND	ND	ND	ND	ND	ND	ND	ND
2-Ethylthiophene	ND	ND	ND	ND	ND	ND	ND	ND
2,5-Dimethylthiophene	ND	ND	ND	ND	ND	ND	ND	ND
Diethyl Disulfide	ND	ND	ND	ND	ND	ND	ND	ND

ND non-detect

Table B-3. Digested Sludge VFA Results.

Days of Operation	Mesophilic Digester (D-9)						Thermophilic Digester (D-11)					
	Individual Volatile Fatty Acids (mg/L)						Individual Volatile Fatty Acids (mg/L)					
	Acetic	Propionic	Butyric	Isobutyric	Isovaleric	Valeric	Acetic	Propionic	Butyric	Isobutyric	Isovaleric	Valeric
14	100	25	25	25	25	25	100	25	25	25	25	25
21	100	25	25	25	25	25	100	25	25	25	25	25
28	100	25	25	25	25	25	100	25	25	25	25	25
35	100	25	25	25	25	25	100	25	25	25	25	25
43	100	25	25	25	25	25	100	25	25	25	25	25
49	100	25	25	25	25	25	170	25	25	25	25	25
56	100	25	25	25	25	25	450	78	25	25	26	25
63	290	29	25	25	25	25	240	25	25	25	25	25
70	250	280	25	25	25	25	130	25	25	25	25	25
77	100	25	25	25	25	25	360	72	25	25	25	25
84							430	240	25	25	25	25
91	100	25	25	25	25	25	210	150	25	25	25	25
98	100	25	25	25	25	25	240	31	25	25	25	25
105	100	25	25	25	25	25	230	25	25	25	25	25
112	100	25	25	25	25	25	100	25	25	25	25	25
119	100	25	25	25	25	25	120	25	25	25	25	25
126	100	25	25	25	25	25	100	25	25	25	25	25
133	100	25	25	25	25	25	100	37	25	25	25	25
140	100	25	25	25	25	25	100	35	25	25	25	25
147	100	25	25	25	25	25	130	25	25	25	25	25
154							100	25	25	25	25	25
161	100	25	25	25	25	25	110	25	25	25	25	25
AVG	117	38	25	25	25	25	174	46	25	25	25	25
MAX	290	280	25	25	25	25	450	240	25	25	26	25
MIN	100	25	25	25	25	25	100	25	25	25	25	25

APPENDIX C

UNITED STATES PATENT: PATHOGEN DESTRUCTION PROCESS FOR BIOSOLIDS

US006447683B1

United States Patent

(12) United States Patent
Gubb et al.
(10) Patent No.: US 6,447,683 B1
(45) Date of Patent: Sep. 10, 2002

(54) **PATHOGEN DESTRUCTION PROCESS FOR BIOSOLIDS**

(75) Inventors: **Donald Michael Douglas Gubb**, Oakland; **David Isaac Jenkins**, Kensingston, both of CA (US); **Sambhunath Ghosh**, Salt Lake City, UT (US); **John Matthew Hake**, Oakland, CA (US); **Carlos De Leon**, Berkeley, CA (US); **David Robert Williams**, Walnut Creek, CA (US)

(73) Assignee: **Water Environment Research Foundation**, Alexandria, VA (US)

(*) Notice: Subject to any disclaimer, the term of this patent is extended or adjusted under 35 U.S.C. 154(b) by 11 days.

(21) Appl. No.: **09/685,458**

(22) Filed: **Oct. 9, 2000**

(51) **Int. Cl.**[7] **C02F 3/00**

(52) **U.S. Cl.** **210/613**

(58) **Field of Search** 210/613, 601

(56) **References Cited**

U.S. PATENT DOCUMENTS

5,228,995 A	*	7/1993	Stover	210/150
5,525,228 A		6/1996	Dague et al.	
5,525,229 A	*	6/1996	Shih	210/177
5,624,565 A		4/1997	Lefevre et al.	
5,630,942 A	*	5/1997	Steiner	210/603
5,746,919 A		5/1998	Dague et al.	
5,900,150 A		5/1999	Fayoux et al.	

OTHER PUBLICATIONS

Aitken, Michael D., et al., Another look at thermophilic anaerobic digestion of wastewater sludge; Water Environment Research; vol. 64; Nov.–Dec. 1992; pp915–919.

Berg, Gerald, et al., Destruction by Anaerobic Mesophilic and thermophilic digestion of viruses and indicator bacteria indigenous to domestic sludges; Applied and Environmental Microbiology; vol. 39, #2; Feb. 1980; pp361–368.

Buhr, H.O. et al., The thermophilic anaerobic digestion process; Water Resource; vol. 2, 1977; pp129–143.

Farrell, Joseph B., et al., Influence of feeding procedures on microbial reductions and performance of anaerobic digestion; Journal WPCF; vol. 60, No. 5; May, 1988; pp635–644.

Gabb, Donald M.D. et al., Pathogen destruction efficiency in high temperature anaerobic digestion; WEFTEC 1999; Oct. 10, 1999; pp1–12.

(List continued on next page.)

Primary Examiner—Chester T. Barry
(74) *Attorney, Agent, or Firm*—Robert D. Varitz, P.C.

(57) **ABSTRACT**

A method of treating a waste stream comprising feeding a waste stream into a reactor constructed and arranged for a fast start and maintained in a thermophilic temperature regime of between about 50° C. and 62° C. for a HRT of less than or equal to 48 hours, wherein the reactor contents are maintained at a pH less than or equal to 6.5; drawing a portion of the contents of the reactor and feeding the drawn contents into a second reactor which is maintained in a mesophilic temperature regime of between about 28° C. to 38° C. for a HRT of at least thirteen days; and replacing the contents drawn from the first reactor with more waste from the waste stream. Alternately, the waste stream may be initially treated in a mesophilic reactor, followed by treatment in a thermophilic reactor, or a single stage thermophilic reactor may be used.

7 Claims, 3 Drawing Sheets

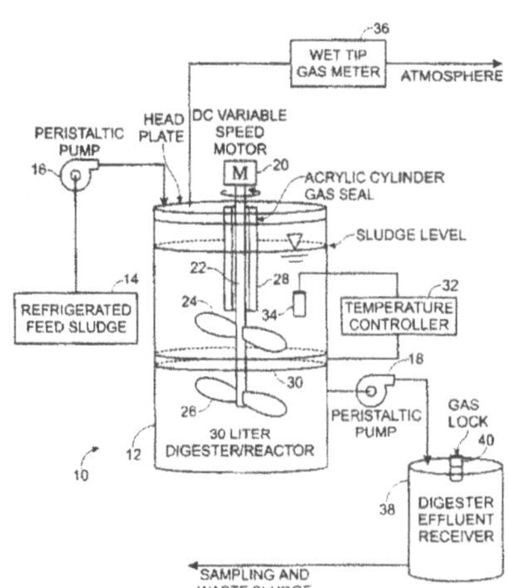

OTHER PUBLICATIONS

Garber, William F., Operating experience with thermophilic anaerobic digestion; Journal WPCF; vol. 54, No. 8; Aug. 1982; pp1170–1175.

Garber, William F., Certain aspects of anaerobic digestion of wastewater solids in the thermophilic range at the Hyperion treatment plant; Prog. Wa. Tech; vol. 8, No. 8; 1977; pp401–406.

Ghosh, Sambhunath, et al., Pilot– and full–scale two–phase anaerobic digestion of municipal sludge; Water Environment Research; vol. 67, No. 2; Mar.–Apr. 1995; pp206–214.

Ghosh, Sambhunath, et al., The Aciment® Process: An innovative approach to biogasification of municipal sludge; Proceeding of the First Biomass Conference; 1993; pp759–769.

Ghosh, Sambhunath, et al., Improved sludge gasification by two–phase anaerobic digestion; J. Envir. Eng., vol. 113, No. 6, Dec. 1987; pp1265–1284.

Ghosh, Sambhunath, et al., Stabilization of Sewage sludge by two–phase anaerobic digestion; EPA/600/S2–87; Sep. 1987; pp1–8.

Ghosh, Sambhunath, et al., Comparative studies of temperature effects on single–stage and two–phase anaerobic digestion; Biotech & Bioeng Symp No. 17; 1986; pp365–377.

Ghosh, Sambhunath, et al., Anaerobic acidogenesis of wastewaster sludge; Journal WPCF; vol. 47, No. 1; Jan. 1975; pp30–45.

Han, Yue et al., Laboratory studies on the temperature phased anaerobic digestion of domestic primary sludge; Water Environment Research; vol. 59, No. 6; Sep.–Oct. 1997; pp1139–1143.

Han, Yue, et al., Temperature–phased anaerobic digestion of wastewater sludges; Water Science Tech; vol. 36, Nos. 6–7; 1997; pp367–374.

Huyard, Alain et al., A challenge for the two phase anaerobic digestion: To produce Class A biosolids and meet PFRP Equivalency; WEFTEC 98; Oct. 1998; pp1–11.

Huyard, Alain et al., Spiking a two phase anaerobic digestion pilor plant: An innovative approach to abtain a P.F.R.P. Eqivalency recommendation; WEFTEC 99; Oct. 1999; pp1–8.

Lee, Kun M., et al., Destruction of enteric bacteria and viruses during two–phase digestion; J. Water Pollution Control; vol. 61, No. 8; Aug. 1989; pp1421–1429.

Mitsdörffer, R., et al, Stabilization and disinfection of sewage sludge by two–stage anaerobic thermophilic/mesophilic digestion; Wat. Sci. Tech; vol. 22, No. 7/8; 1990; pp289–290.

Popova, N.M., et al., The present state of purification of twon sewage and the trend in research work in the city of Moscow; Advances in Water Pollution; vol. 2, 1964; pp97–118.

Stern, Gerald, et al., Sludge disinfection techniques; EPA, Cincinnati, Ohio; 1977; pp142–148.

Watanabe, Haruki, et al., Inactivation of pathogenic bacteria under mesophilic and thermophilic conditions; Water Sci. Tech; vol. 36, No. 6–7; 1997; pp25–32.

Zinder, Stephen, Thermophilic waste treatment systems; *Thermophiles*, John Wiley & Sons, New York; 1986; pp257–277.

* cited by examiner

Fig. 1

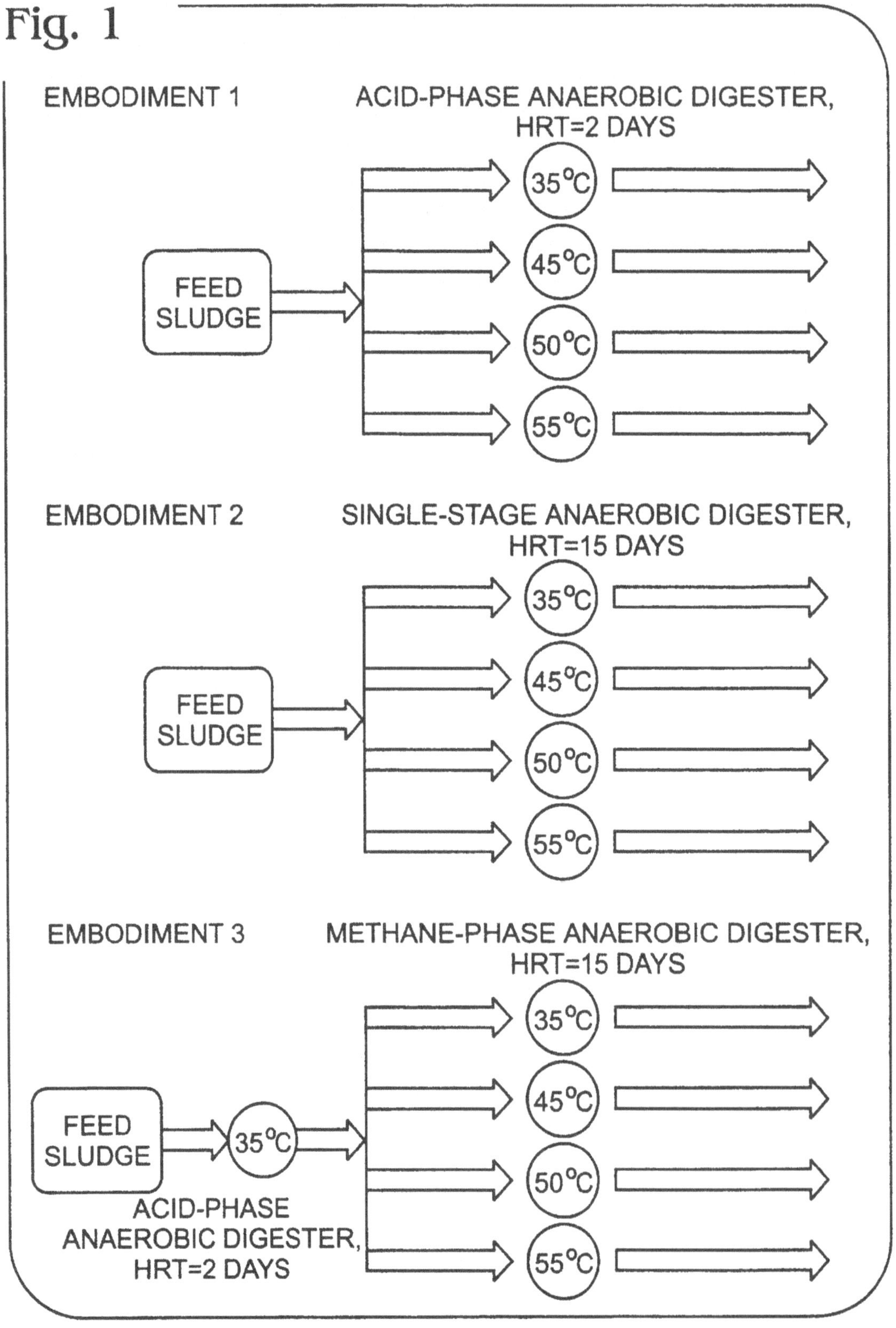
EMBODIMENT 1
ACID-PHASE ANAEROBIC DIGESTER, HRT=2 DAYS
FEED SLUDGE
35°C
45°C
50°C
55°C
EMBODIMENT 2
SINGLE-STAGE ANAEROBIC DIGESTER, HRT=15 DAYS
FEED SLUDGE
35°C
45°C
50°C
55°C
EMBODIMENT 3
METHANE-PHASE ANAEROBIC DIGESTER, HRT=15 DAYS
FEED SLUDGE
35°C
ACID-PHASE ANAEROBIC DIGESTER, HRT=2 DAYS
35°C
45°C
50°C
55°C

Fig. 2

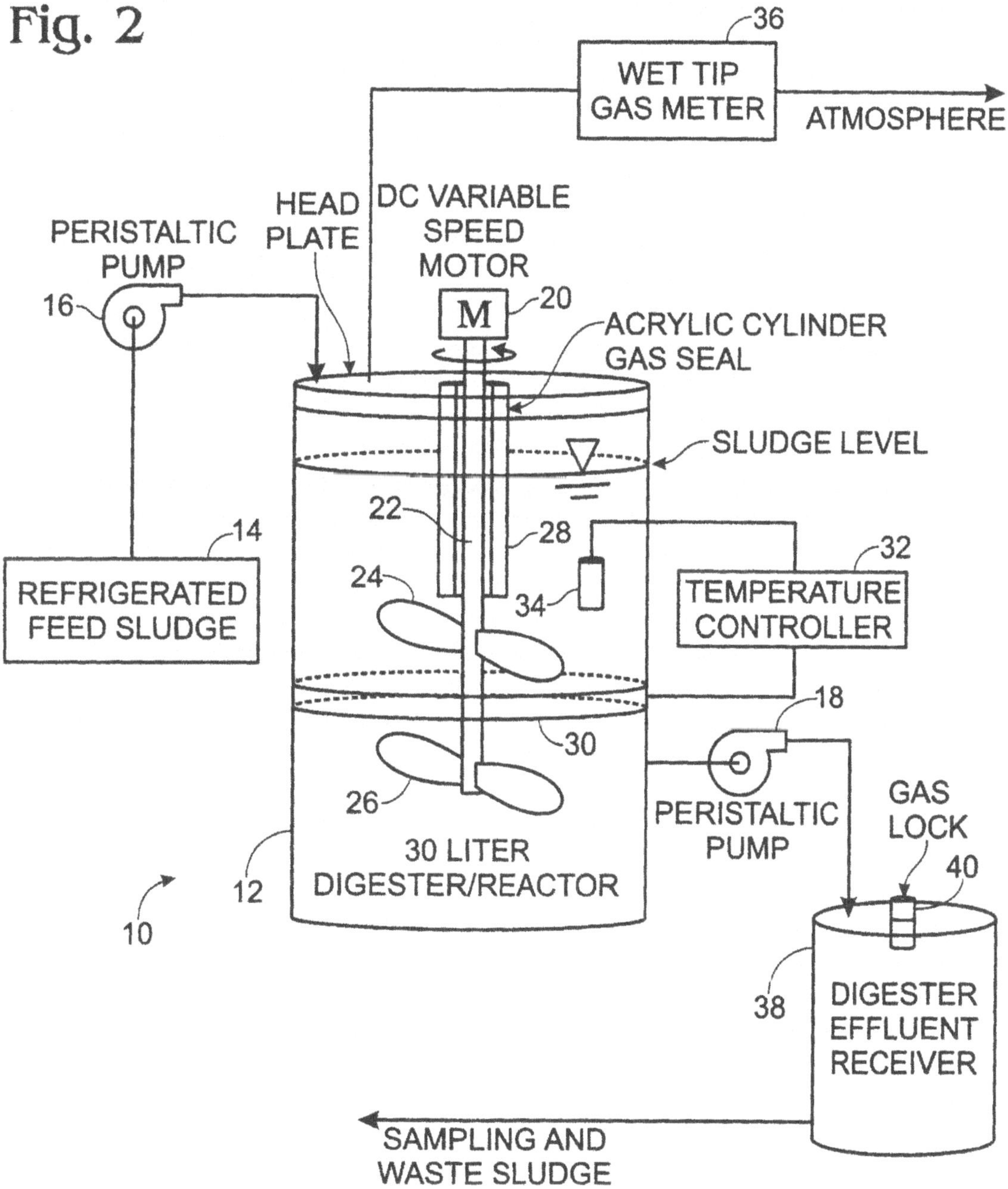
36
WET TIP
GAS METER
ATMOSPHERE
HEAD
PLATE
DC VARIABLE
SPEED
MOTOR
PERISTALTIC
PUMP
16
M
20
ACRYLIC CYLINDER
GAS SEAL
SLUDGE LEVEL
22
28
14
32
REFRIGERATED
FEED SLUDGE
24
34
TEMPERATURE
CONTROLLER
18
30
26
PERISTALTIC
PUMP
GAS
LOCK
40
30 LITER
DIGESTER/REACTOR
12
10
38
DIGESTER
EFFLUENT
RECEIVER
SAMPLING AND
WASTE SLUDGE

Fig. 3

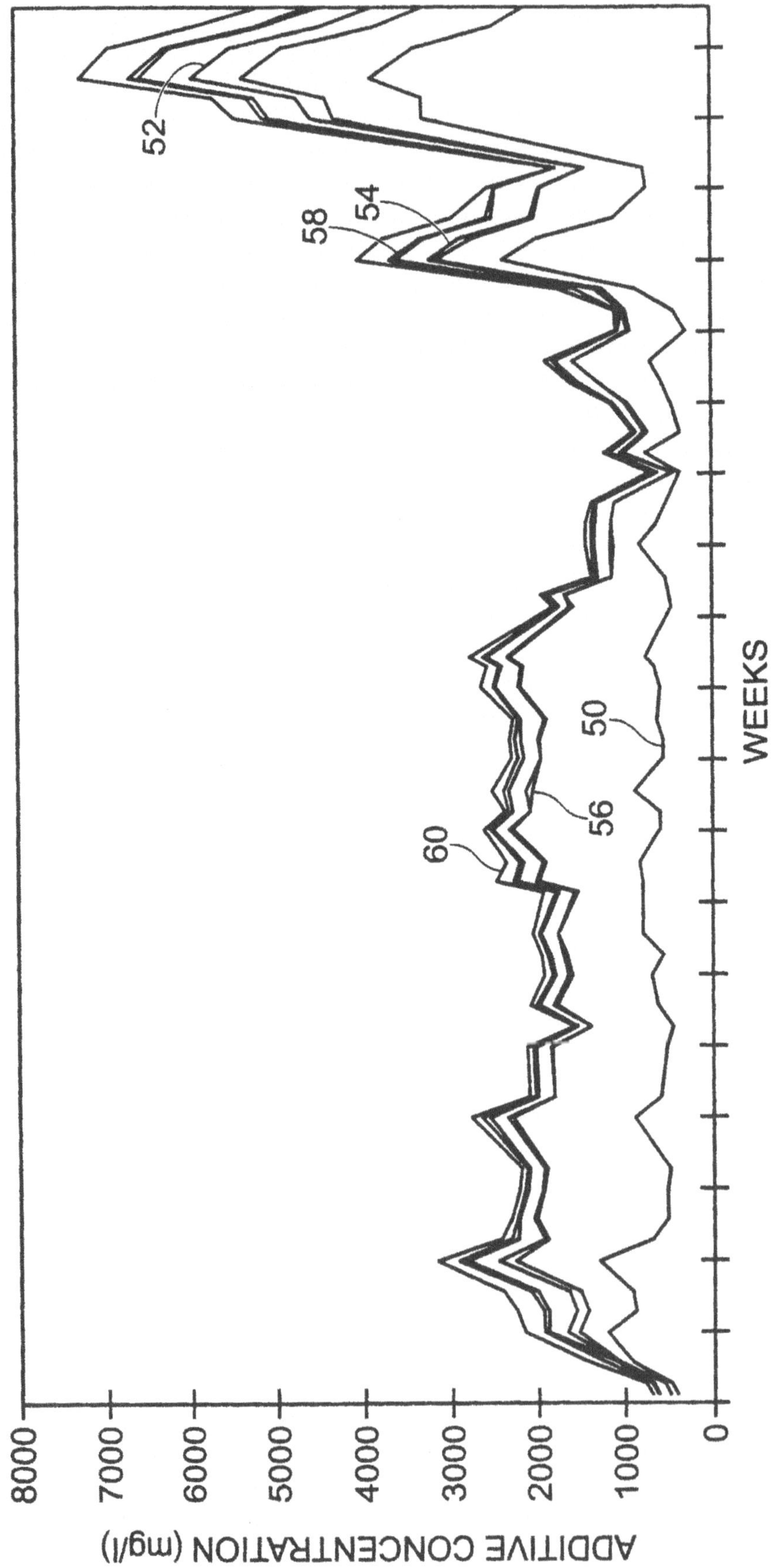
ADDITIVE CONCENTRATION (mg/l)
8000
7000
6000
5000
4000
3000
2000
1000
0
WEEKS
50
52
54
56
58
60

PATHOGEN DESTRUCTION PROCESS FOR BIOSOLIDS

SPONSORSHIP

This work was supported by a grant from the Water Environment Research Foundation (Project No. 97-REM-2). The inventors acknowledge the assistance provided by Bio Vir Laboratories and the East Bay Municipal Utility District (EBMUD) laboratory staff.

FIELD OF THE INVENTION

This invention relates to the processing of wastewater, and particularly to a method for providing treatment to achieve Class A biosolids.

BACKGROUND OF THE INVENTION

Land application of treated wastewater sludge (biosolids) is controversial because the biosolids therein potentially contain human pathogens, i.e., some counties in California have banned the application of biosolids which meet the Environmental Protection Agency's (EPA) Class B pathogen reduction standards, as set forth in 40 CFR 503, which require a fecal coliform level of less than two million most-probable-number (MPN) per gram of total solids, but not Class A pathogens standards per 40 CFR 503. Wastewater treatment agency managers are concerned about the future of biosolids land applications in their states. EPA's Class A pathogen standards require the virtual elimination of pathogens in biosolids. In accordance with 40 CFR 503.32 (a)(5), Class A—alternate 3 and 40 CFR 503.32(a)(6)—alternate 4, EPA Class A pathogen requirements are met in biosolids when fecal coliform densities are less than 1,000 MPN per gram total solids; or when Salmonella densities are less than 3 MPN per four grams total solids. Additionally, enteric virus must be less than 1 plaque-forming unit per four grams of total solids, and helminth ova is less than one viable helminth ova per four grams of total solids. Meeting Class A standards will significantly increase the opportunity for biosolids recycling, however, known processes which achieve Class A pathogen levels in biosolids are generally cost prohibitive. The provision of a low-cost method of biosolids treatment, which will meet Class A standards will offer additional biosolids management options Known thermophilic anaerobic digestion processes are not classified as a Process to Further Reduce Pathogens (PFRP) under 40 CFR Part 503.

Anaerobic digestion has been one of the most widely used processes for the stabilization of primary and secondary sludges produced at municipal wastewater treatment facilities. The majority of applications of anaerobic digestion to wastewater sludges have been in the mesophilic temperature range, from 35° C. to 40° C. (95° F. to 104° F.). Anaerobic sludge digestion in the thermophilic temperature range from 45° C. to 65° C. (113° F. to 149° F.) has been practiced to only a limited extent.

The limited use of anaerobic digestion at temperatures above the mesophilic range is due to higher energy requirements to obtain the higher thermophilic temperature, poor process stability, increased odor, and lower quality supernatant (filtrate/centrate). The advantages of thermophilic anaerobic digestion over mesophilic anaerobic digestion have accrued from increased stabilization and methane production rates, and from improvements in sludge dewatering properties. Since the advent of the 40 CFR Part 503 Regulations, more studies have focused on the destruction of pathogenic organisms.

Thermophilic anaerobic digestion has an advantage of improving pathogen destruction, and has the potential to meet the pathogen quality requirements of EPA's Class A biosolids. While the economic disadvantages of thermophilic anaerobic digestion has outweighed the advantages of the process, the implementation of 40 CFR Part 503 and the use of a two-stage digestion system, having a thermophilic or mesophilic first-stage and a mesophilic or thermophilic second-stage, may negate the economic disadvantage.

Recent studies have introduced some improvement to the stability and performance of the thermophilic anaerobic digestion process. These improvements include:

Fast start-up method. The fast start-up method produces a more robust thermophilic culture that is relatively tolerant of temperature fluctuations and increases process stability.

Draw/fill operation. The draw/fill cycle improves pathogen removal by reducing the potential for short-circuiting.

Multiple-stage digestion. Multiple-stage reactors reduce the potential for short-circuiting, wherein flow exits the reactor before it should, provide a low pH environment in the first stage if methanogenesis is minimized or eliminated in this stage, which apparently enhances pathogen reduction; and, if the second-stage reactor is mesophilic, produce a higher quality sludge in terms of odor and supernatant.

Research has shown that pathogen destruction in municipal sludge digestion follows a typical time/temperature relationship, wherein higher temperatures require shorter exposure times for pathogen destruction. Data has been collected demonstrating survival rates of various pathogens in municipal sludge digestion. These data suggest that thermophilic digestion achieves pathogenic bacteria reduction rates about two orders of magnitude higher than mesophilic digestion, and may meet the pathogenic bacteria levels required for Class A sludge. EPA studies and full-scale research results indicate enhanced pathogen destruction in acid-phase digestion.

U.S. Pat. No. 5,525,228 to Dague et al., for Temperature-phased anaerobic waste treatment process, granted Jun. 11, 1996, describes a process having a long term hydraulic retention time (HRT) and which operates in a temperature range sufficient to Pasteurize the waste.

U.S. Pat. No. 5,624,565 to Lefevre et al., for Method of regulating aeration in a basin for biological treatment of Waste water, granted Apr. 29, 1997, describes a method which requires elimination of carbon in aerobiosis, nitrification and denitrification.

U.S. Pat. No. 5,746,919 to Dague et al., for Temperature-phased anaerobic waste treatment process, granted May 5, 1998, is a continuation of the '228 patent.

U.S. Pat. No. 5,900,150 to Fayoux et al., for Method of purifying waste water biologically, granted May 4, 1999, describes a treatment method which is conducted at temperatures between 25° C. and 40° C.

SUMMARY OF THE INVENTION

A method of treating a waste stream comprising feeding a waste stream into a reactor constructed and arranged for a fast start and maintained in a thermophilic temperature regime of between about 50° C. and 62° C., for a HRT of less than or equal to 48 hours, wherein the reactor contents are maintained at a pH of 6.5 or less; drawing a portion of the reactor contents and feeding into a second reactor which is maintained in a mesophilic temperature regime of between about 28° C. to 38° C. for a HRT of at least thirteen days; and replacing the volume of sludge drawn from the thermophilic reactor with more waste from the waste stream. Alternately, the waste stream may be initially treated in a mesophilic reactor, followed by treatment in a thermophilic reactor, or a single stage thermophilic reactor may be used.

An object of the invention is to meet Class A requirements with a thermophilic anaerobic digestion processes.

Another object of the invention is to provide a digestion process for fecal coliform, Salmonella, enteric virus and helminth ova reduction.

A further object of the invention is to provide a treatment method which will produce less methane than do prior art methods.

Another object of the invention is to provide a treatment method which will produce biosolids which are substantially pathogen-free and are suitable for application with respect to pathogen densities where ever commercial fertilizers or soil conditioners are used.

This summary and objectives of the invention are provided to enable quick comprehension of the nature of the invention. A more thorough understanding of the invention may be obtained by reference to the following detailed description of the preferred embodiment of the invention in connection with the drawings.

BRIEF DESCRIPTION OF THE DRAWINGS

FIG. **1** is a flow chart depicting the various processes used to demonstrate the method of the invention.

FIG. **2** is a schematic representation of the structure used to demonstrate the method of the invention.

FIG. **3** depicts the distribution of fatty acids in the second embodiment of the invention at Thermo 55.

DETAILED DESCRIPTION OF THE PREFERRED EMBODIMENT

This invention includes three embodiments of various thermophilic anaerobic digestion processes for disinfecting sludge, in some instances, to meet the EPA's Class A pathogen requirements, as set forth in 40 CFR Part 503, the requirement of which have been previously identified. Three embodiments of the invention, schematically represented in FIG. **1**, are assessed for their ability to consistently meet fecal coliform requirements, and include: (1) a two-stage, thermophilic acid phase/mesophilic methane phase; (2) a single-stage thermophilic; and (3) a two-stage, mesophilic acid phase/thermophilic methane phase. Each process is evaluated at three different thermophilic temperatures, i.e., 45° C., 50° C., and 55° C., against a mesophilic control at 35° C. As used herein, a mesophilic temperature range includes temperatures between 35° C. and 40° C., while a thermophilic temperature range includes temperatures between 40° C. and 70° C. Temperatures above 70° C. are in a range which will pasteurize an organic material in thirty minutes per 40 CFR 503, Appendix B.7. Pasteurization is a physical, rather than a biological process. What is significant about the three embodiments of the invention is that the initial HRT is of short duration, and introduction of a waste stream into a thermophilic reactor is accompanied by a rapid start up, i.e., the contents of the reactor chamber are quickly brought up to the thermophilic temperature regime.

Experimental data using the methods of the invention was gathered in three phases. In Phase 1, the three embodiments of the invention were screened at bench-scale. Because the acid stage of the first embodiment was expected to provide significantly greater disinfection power than the mesophilic methane stage, only the acid stage was operated for evaluation purposes in Phase 1. Each embodiment was examined at or about three different thermophilic temperatures: 45° C., 50° C., and 55° C., referred to herein as Thermo 45, Thermo 50 and Thermo 55, respectively, against a mesophilic control at 35° C., referred to herein as Meso 35. In Phase 2, the better performing embodiments of Phase 1 were examined in more detail at bench-scale over a longer time period for assessment of process stability and actual pathogen destruction using feed sludges seeded with pathogens of concern. Phase 3 is a further continuation of the Phase 2 assessment at full-scale.

The method of the invention was tested, as shown in FIG. **2**, generally at **10**, using laboratory-scale Plexiglas reactors **12** (digesters) with an operating capacity of approximately 30 liters. Peristaltic pumps **16**, **18** control reactor influent and effluent flows, respectively.

The reactors were brought to thermophilic temperatures rapidly from a mesophilic seed sludge and operated in a draw/fill mode. As used herein, "rapidly" means that the seed sludge, at an initial temperature of about 35° C., was brought up to the thermophilic temperature regime in less than one day. At corresponding temperatures, single stage thermophilic processes (2) demonstrated greater fecal coliform removals than the first, thermophilic acid phase, stage of the two-stage process (1). Temperatures greater than 50° C. for embodiment 1, having a 2-day residence time, or 45° C. for embodiment 2, having a 15-day residence time, produced average effluent fecal coliform concentrations less than 1,000 MPN/g.TS (most probably number of pathogen/gram of total solids).

The reactor contents are mixed by an externally-mounted variable-speed DC motor **20** that turns a stainless steel shaft **22** and two 4-inch diameter propellers **24**, **26**. The mixing system is sealed and anaerobic conditions are maintained by a small tube **28** that covers the stainless steel shaft. Each reactor was heated by a 520-watt heat tape **30**, such as Cole Parmer, E-36110-08. To avoid direct contact between the Plexiglas reactor walls and heat tape the exterior vertical area of the cylinders was covered with aluminum foil (not shown), which also distributes heat more evenly. Digester temperature is controlled by a proportional-response temperature controller **32**, such as Dynasense Model 02149-20, which receives a signal from an 18-inch immersible RTD probe **34**, such as Cole Parmer, E-08 117-74. As the temperature controller does not include a temperature display, temperatures are obtained from an external bimetal thermometer (not shown), and the heat tape output adjusted accordingly. Gas production was measured using a wet-tip gas meter **36**, as supplied by Wet Tip Gas Meter Co. Reactor effluent was stored in a container **38** with a gas seal **40** that allowed gas to escape but prevented air from entering.

The various hydraulic residence times (HRT), two days for embodiment 1 and fifteen days for embodiment 2, were achieved by adjusting flow rates and pumping intervals. The 2-day HRT of the reactors was achieved for part of the test period by cycling one-fourth of the reactor volume twice a day. This feeding regime approximated batch behavior to maximize fecal coliform removal, however, a semi-continuous flow regime, having hourly feed cycles, was adopted for another part of the test period to better control reactor levels. Feed cycles for embodiment 2 were six hours apart. Feeding regimes include the many variations from batch-feed, intermittent fee and continuous-feed.

Reactor feed, the waste stream, was a mixture of the a main wastewater treatment plant's primary sludge (PS) and thickened waste activated sludge (TWAS). The PS was filtered through a 5×5 wire mesh, having a 63.2% open area, and mixed in a 1:1 volume ratio with TWAS. The feed was kept in a refrigerator at approximately 4° C. and was mixed using a submerged pump and recirculation loop prior to feeding. For embodiment 1, the feed lines were immersed in a pre-heating water bath at 35° C. to avoid large reactor temperature fluctuations. Pre-heating was not required for embodiment 2 because of the smaller feed volumes and longer feed cycle periods.

All tests were performed in accordance with procedures detailed in the Standard Methods for the Examination of Water and Wastewater. Reactor feed and effluent sludges were analyzed for a number of parameters to monitor digester performance. The average and standard deviation for each parameter were calculated for embodiment 1, and are shown in Table 1: "General Characteristics of Reactor Behavior" and for embodiment 2 in Table 2 "General Characteristics of Reactor Behavior." Digester performance, particularly for fecal coliform removal, is assessed as a function of temperature and hydraulic residence time.

TABLE 1

General characteristics of Reactor Behavior for Embodiment 1

Parameter	#of Samples		Reactor Operating Temperature 35	45	50	55	Feed
Residence time (days)*	56	average	3.3	2.4	2.8	2.7	
		st. dev.	46	09	22	21	
Temperature (° C.)*	112	average	35	45	52	55	
		st. dev.	2	2	3	3	
Gas production (l/hr)**	57	average	0.5	0.0	0.8	0.2	
		st. dev.	0.3	0.1	0.6	0.5	
Methane in gas produced (%)***	5	average	23.6	24.2	32.6	33.8	
		st. dev.	29	38	51	70	
pH***	58	average	584	607	658	665	5.88
		st. dev.	0.20	0.12	0.17	0.16	0.13
Total Alkalinity (mg $CaCO_3$/l)*	26	average	3415	4162	4942	5085	
		st. dev.	502	461	544	520	
Volatile Acids (mg acetic acid/l)*	26	average	6681	7477	7265	7462	
		st. dev.	747	893	691	863	
VFA/TALK ratio	26	average	1.98	1.81	1.48	1.47	
(mg acetic acid/mg $CaCO_3$)*		st. dev.	0.23	0.21	0.15	0.13	
Acetic Acid (mg/l)****	7	average	2800	3471	3243	3571	1729
		st. dev.	351	315	447	571	180
Propionic Acid (mg/l)****	7	average	1757	1429	1514	1714	871
		st. dev.	336	214	279	248	176
Butyric Acid (mg/l)****	7	average	1250	1543	1486	1343	730
		st. dev.	210	140	212	244	172
Isobutyric Acid (mg/l)****	7	average	1189	2246	1430	729	380
		st. dev.	1770	2675	1928	87	23
Valeric Acid (mg/l)****	7	average	473	426	194	187	187
		st. dev.	72	40	20	18	33
Isovaleric Acid (mg/l)****	7	average	533	703	801	849	229
		st. dev.	67	72	98	126	45
Total Solids (%)*	26	average	3.5	3.6	3.2	3.3	4.3
		st. dev.	0.4	1.2	0.3	0.4	0.8
Volatile Solids (%)*	26	average	2.8	2.6	2.5	2.5	3.5
		st. dev.	0.4	0.3	0.3	0.3	0.7
Vol. Sol. Removal efficiency (%)*	26	average	169%	22.8%	20.9%	21.5%	
		st. dev.	11.1%	21.1%	12.2%	10.9%	
Fecal Coliform (MPN/g TS)*	26	average	15196	4445	905	449	309846
		st. dev.	19661	5008	1538	1133	600687
Fecal Log reduction*	26	average	1.24	1.74	3.07	3.60	
		st. dev.	0.65	0.61	1.11	1.05	

*58 Days
**57 Days
***31 Days
****39 Days

TABLE 2

General characteristics of Reactor Behavior for Embodiment 2

Parameter	#of Samples		Reactor Operating Temperature 35	45	50	55	Feed
Residence time (days)*	112	average	16.3	15.7	14.2	15.1	
		st. dev.	4	4	4	4	
Temperature (° C.)*	222	average	34	45	51	56	
		st. dev.	2	2	3	12	
Gas production (l/hr)**	113	average	1.5	2.0	1.5	1.5	
		st. dev.	0.6	0.7	0.6	0.7	

TABLE 2-continued

General characteristics of Reactor Behavior for Embodiment 2

Parameter	#of Samples		Reactor Operating Temperature 35	45	50	55	Feed
Methane in gas produced (%)***	20	average	61.4	58.1	54.1	55.0	
		st. dev.	8.3	11.5	12.1	9.4	
pH***	113	average	7.95	8.05	8.08	8.07	5.92
		st. dev.	0.18	0.18	0.15	0.25	0.22
Total Alkalinity (mg $CaCO_3$/l)*	67	average	6769	6803	6661	6518	
		st. dev.	437	475	386	534	
Voltile Acids (mg acetic acid/l)*	67	average	270	488	1525	2199	
		st. dev.	260	465	733	1129	
VFA/TALK ratio	67	average	0.04	0.07	0.23	0.34	
(mg acetic acid/mg $CaCO_3$)*		st. dev.	0.04	0.08	0.11	0.21	
Acetic Acid (mg/l)****	58	average	280	460	578	982	1611
		st. dev.	167	330	283	834	692
Propionic Acid (mg/)****	58	average	61	95	796	948	786
		st. dev.	95	118	457	447	438
Butyric Acid (mg/l)****	58	average	28	27	40	89	602
		st. dev.	14	8	37	147	406
Isobutyric Acid (mg/l)****	58	average	67	93	156	252	298
		st. dev.	25	31	64	148	118
Valteric Acid (mg/l)***	58	average	26	25	40	45	98
		st. dev.	4	1	25	19	65
Isovaleric Acid (mg/l)****	58	average	28	31	76	158	188
		st. dev.	16	19	74	174	134
Total Solids (%)*	67	average	2.2	2.2	2.2	2.2	4.2
		st. dev.	0.3	0.3	0.3	0.4	1.0
Volatile Solids (%)*	67	average	1.4	1.4	1.5	1.5	3.2
		st. dev.	0.2	0.2	0.2	0.2	0.8
Vol. Sol. Removal efficiency (%)*	67	average	42.8%	44.1%	40.2%	38.5%	
		st. dev.	17.0%	15.9%	15.8%	17.8%	
Fecal Coliforms (MPN/g TS)*	58	average	1100	240	110	43	120000
		st. dev.	1200	770	320	170	150000
Fecal Log reduction	58	average	1.94	3.07	3.69	4.16	
		st. dev.	0.67	0.86	0.89	0.92	

*58 Days
**57 Days
***31 Days
****39 Days

Embodiment 1

Two-Stage Thermophilic Acid Phase/Mesophilic Methane Phase Digestion

Embodiment 1 was evaluated using only the first stage (thermophilic acid phase) of the two-stage process because it was expected that this is where the greatest pathogen destruction would occur. Reactor temperatures appeared to have a proportional relationship with total alkalinity (TALK), pH, methane production, fecal coliform removal, and an inversely proportional relationship with volatile fatty acids (VFA)/TALK ratio, as shown in Table 1. TALK and pH levels were interrelated with the proportion of methane in gas production —alkalinity is a by-product of methanogenesis in that it results in the consumption volatile of fatty acids. As expected, all of the reactors in this process had high VFA concentrations because of the short HRT, but it appeared there was no direct relationship between total VFA concentration and temperature. Because of higher alkalinities of the higher-temperature reactors, VFA/TALK ratios exhibited an inverse relationship with temperature. The preferred temperatures in the thermophilic regime are in a range of about 50° C. to 62° C.

Because of the perceived impact of feeding regime, i.e., semi-batch vs. semi-continuous, on fecal coliform removal, a statistical analysis was performed to determine significance. Semi-batch feeding showed slightly higher removals at 35° C. and 45° C., and semi-continuous feeding showed slightly higher removals at 50° C. and 55° C. An Analysis of Variance (ANOVA) test and a Tukey comparison showed that these differences were not statistically significant. However, the embodiment 1 data presented here includes only the semi-continuous feeding results for consistency and comparison to embodiment 2.

Fecal coliform log removal rates increased noticeably with temperature. At 50° C. and higher, average fecal coliform concentrations in the digested sludge were less than 1,000 MPN/g. TS, the fecal coliform requirement for Class A sludge. The 50° C. and 55° C. thermophilic reactors achieved removals greater than three logs. As previously noted, the fecal coliform removal was only for the first stage of a two-stage process. Inclusion of the second stage, mesophilic methane phase, of the first embodiment would likely result in removals which are higher in all temperatures regimes. Preferred mesophilic temperature regime is in a temperature range of about 28° C. to 38° C.

Embodiment 2

Single-Stage Thermophilic Digestion

As shown in Tables 1 and 2, pH and TALK values for each reactor were substantially higher for embodiment 2 than embodiment 1. VFA/TALK ratios were also proportionately greater at higher temperatures. Propionic acid appeared to accumulate in the higher-temperature reactors, 50° C. and 55° C., further increasing VFA/TALK values. VFA/TALKs at 50° C. and 55° C. were much greater than 0.1 mg VFA as acetic acid/mg TALK as $CaCO_2$, a standard benchmark for conventional mesophilic digesters.

During the initial period of embodiment 2 testing, much lower fecal coliform concentrations were measured in the reactor feed than were measured during the embodiment 1 evaluation. To determine if this decrease was significantly different from the expected values, a statistical Mann-Whitney test was performed on the feed sludge fecal concentrations. ANOVA was also performed but the data did not have a normal distribution. The results showed that fecal coliform concentrations in this initial period, log Fecals=4.3606 xxx 0.7922, were significantly less than those that followed, log Fecals=5.1840 xxx 0.3977. The reduced concentrations in the beginning of the process were related to decreased frequency of fresh sludge feed. Subsequently, the frequency was increased, to daily, and the feed sludge fecal coliform concentrations recovered to their former levels. However, during this period the reduced concentration of feed sludge fecal Coliforms also decreased the removal rates, especially for the lower-temperature reactors. The feed sludge fecal coliform data at reduced levels is included in the embodiment 2 data in Table 2.

Toward the end of embodiment 2 testing, the VFA/TALK ratio for the 55° C. reactor increased to almost 1.0 mg acetic acid/mg CaCO3. As expected, this condition was accompanied by a drop in gas production and a lower methane concentration, although the pH remained relatively constant and the TALK concentration high. It is possible that propionic acid accumulation, as depicted in FIG. **3**, Distribution of Fatty Acids in Thermo 55, embodiment 2", was not easily metabolized by the methanogenic bacterial populations under thermophilic conditions. FIG. **3** depicts levels of fatty acids over approximately twenty weeks of testing, corresponding to the data in Table 2. Trace **50** represents acetic acid, trace **52** represents butyric acid, trace **54** represents valeric acid, trace **56** represents propionic acid, trace **58** represents isobutyric acid, and trace **60** represents isovaleric acid. The actual operating temperature of Thermo 55 was 62° C. At this temperature, it appears that methanogenic activity is significantly impaired.

Fecal coliform log removal rates increased noticeably with temperature. At 45° C. and higher, embodiment 2 achieved 3–4 log removals of fecal coliform and average fecal coliform concentrations in the digested sludge were less than 1,000 MPN/g. TS, satisfying Class A sludge requirements. For embodiment 2, the preferred temperatures are in the range of 45° C. to 62° C. for the thermophilic reactor.

A summary of the conditions and results for the three embodiments are set forth in Table 3:

TABLE 3

	Temperature						% of time in compliance
Embodiment 1	Nominal	Actual	Mean	Std Dev	n	n < 1000	with Class A
	45° C.	48° C.	3.378	4.515	37	12	32
	50° C.	52° C.	934.57	1516.94	37	27	73
	55° C.	62° C.	339.43	971.74	37	32	86
Embodiment 2	45° C.		243.6	767.78	43	42	98
	50° C.		107.3	319.68	44	42	95
	55° C.		42.7	172.3	44	43	98
Embodiment 3	35° C.		706.9	1614.4	38	33	87
	45° C.		128	367	38	35	92
	50° C.		3.8	6	38	38	100
	55° C.		2.8	4.6	38	38	100

As previously noted, only the first stage thermophilic reactor was tested for embodiment 1 using a two-day HRT. Embodiment 2 at Thermo 45 and 55 and embodiment 3 at Meso 35 followed by Thermo 50, 55 produced significant results, meeting EPA Class A requirements 98% and 100% of the time, respectively. Sludge treated following the method of the invention as set forth in these embodiments is suitable, with respect to pathogen densities, for most applications where commercial fertilizers and soil conditioners are used.

A summary of the performance of the three embodiments is set forth in Table 4:

TABLE 4

	Embodiment 1			Embodiment 2			Embodiment 3			
Temperature (° C.)	48	52	62	45	50	55	35	45	50	55
Pop. Mean < 1000 t calculated (All numbers are −ve)	Fail	0.26	3.76	6.46	18.52	36.85	1.07	14.65	1023.5	1336
degree of freedom		35	35	40	40	40	40	40	40	40

TABLE 4-continued

		Embodiment 1			Embodiment 2			Embodiment 3			
Temperature (° C.)		48	52	62	45	50	55	35	45	50	55
4 @ 0.005 level (All numbers are −ve)			2.724	2.724	2.705	2.705	2.705	2.705	2.705	2.705	2.705
Fails/Succeeds (F/S)		F	F	S	S	S	S	F	S	S	S
Interval	LCL	1356*	255**	0	0	0	0	0	0	1.2	0.8
	UCL	5400*	1614**	834	561	237	113	1449	289	64	4.8
Portion of Population Meeting Class A 90% of the Time											
z calculated		N/A	N/A	N/A	1.73	1.1	1.73	N/A	0.44	2.05	2.05
p-value		N/A	N/A	N/A	.0418	.1357	.0418	N/A	0.33	.0202	.0202
Confidence Level (%)		N/A	N/A	N/A	95.82	86.43	95.82	N/A	67	97.98	97.98

*The 99% confidence interval for embodiment 1 population mean for fecal coliform densities @ 48° C. ranges from 1356 to 5400 MPN, which is well about the Class A limit of 1000 MPN/100 ml.

**At 52° C., this embodiment is a near miss, but should provide consistent compliance with EPA Class A when the mesophilic digester is added as a second stage.

Comparison of Process 1 and 2

Results from embodiment 1 and embodiment 2 testing were compared, and the results presented in Table 5.

TABLE 5

Statistical Similarities Among Embodiments 1 and 2 Average Fecal Coliform Log Concentrations

Reactor	Embodiment 1	Embodiment 2
Meso 35	3.6871 ± 0.6407	2.9448 + 0.5927
Thermo 45	3.1341 ± 0.6765	1.8011 ± 0.9091
Thermo 50	2.0751 ± 1.1216	1.1986 ± 0.8145
Thermo 55	1.6061 ± 1.0182	0.7232 ± 0.6693

A number of conclusions may be made regarding fecal coliform removals:

1. Longer residence times showed greater fecal coliform reductions.
2. A reactor operating at a 2-day HRT and 50° C. had a statistically similar ability to disinfect as a reactor operating at a 15-day HRT and 45° C.
3. A reactor operating at a 2-day HRT and 55° C. had a statistically similar ability to disinfect as a reactor operating at a 15-day HRT and 45° C. or 50° C.
4. Embodiment 2 showed significantly greater removals at thermophilic temperatures, although it must be noted that embodiment 2 had lower feed sludge concentrations and that embodiment 1 used only one stage of a two-stage process. Operation of the systems according to identical conditions for embodiments 1 and 2 would likely produce more similar removal rates.

From the results of embodiments 1 and 2, the following conclusions may be drawn:

1. Average effluent fecal coliform concentrations were less than 1,000 MPN/g. TS, meeting EPA 503 fecal coliform requirements for Class A treatment for embodiment 1 operating at 50° C. and 55° C.; and for embodiment 2 operating at 45° C., 50° C. or 55° C.
2. Embodiment 2 showed greater fecal coliform removals than embodiment 1 removals at the same temperature. However, adding a methanogenic second stage to embodiment 1 and more consistent feed sludge fecal coliform levels may reduce this difference.

Thus, three embodiments of a method for pathogen destruction has been disclosed. The second and third embodiments produce fecal coliform reductions in biosolids to a level sufficient to apply these treated biosolids in a manner similar to that of commercial fertilizers or soil conditioners. It will be appreciated that further variations and modifications thereof may be made within the scope of the invention as defined in the appended claims.

We claim:

1. A method of treating a waste stream comprising:

feeding a waste stream into a thermophilic anaerobic reactor maintained in a thermophilic temperature regime of between about 50° C. and 62° C., for a HRT of less than or equal to 48 hours;

drawing a portion of the contents of the thermophilic reactor and feeding the drawn contents into a mesophilic anaerobic reactor which is maintained in a mesophilic temperature regime of between about 28° C. to 38° C. for a HRT of at least thirteen days; and

replacing the volume of the drawn contents from the thermophilic reactor by feeding the thermophilic reactor with a volume of waste from the waste stream.

2. The method of claim **1** which includes maintaining the reactor contents of the thermophilic reactor at a pH less than or equal to 6.5.

3. The method of claim **1** which includes maintaining the reactor contents of the mesophilic reactor at a pH of at least 6.5.

4. A method of treating a waste stream comprising:

feeding a waste stream into an anaerobic reactor maintained in a thermophilic temperature regime of between about 45° C. and 62° C., for a HRT of at least thirteen days, wherein the reactor contents are maintained at a pH of at least 6.5;

drawing a portion of the reactor contents; and

replacing the volume of the contents removed from the reactor by feeding the reactor with more waste from the waste stream.

5. A method of treating a waste stream comprising:

feeding a waste stream into a mesophilic anaerobic reactor which is maintained in a mesophilic temperature regime of between about 28° C. to 38° C. for a HRT of less than or equal to 48 hours;

drawing a portion of the contents of the mesophilic reactor and feeding into a thermophilic anaerobic reactor maintained in a thermophilic temperature regime of between about 45° C. and 62° C., wherein the HRT in the thermophilic anaerobic reactor is at least thirteen days;

replacing the volume of the contents removed from the mesophilic anaerobic reactor by feeding the mesophilic anaerobic reactor with more waste from the waste stream.

6. The method of claim 5 which includes maintaining the reactor contents of the thermophilic reactor at a pH of at least 6.5.

7. The method of claim 5 which includes maintaining the reactor contents of the mesophilic reactor at a pH less than or equal to 6.5.

* * * * *

PTO/SB/44 (02-01)
Approved for use through 01/31/2004. OMB 0651-0033
U.S. Patent and Trademark Office; U.S. DEPARTMENT OF COMMERCE
Under the Paperwork Reduction Act of 1995, no persons are required to respond to a collection of information unless it displays a valid OMB control number.
(Also Form PTO-1050)

UNITED STATES PATENT AND TRADEMARK OFFICE
CERTIFICATE OF CORRECTION

PATENT NO : 6,447,683 B1

DATED : September 10, 2002

INVENTOR(S) : Donald Michael Douglas Gubb, David Isaac Jenkins, Sambhunath Ghosh, John Matthew Hake, Carlos De Leon, David Robert Williams

It is certified that error appears in the above-identified patent and that said Letters Patent is hereby corrected as shown below:

Donald Michael Douglas Gubb should be changed to:

"Donald Michael Douglas Gray (Gabb)"

MAILING ADDRESS OF SENDER:

Water Environment Research Foundation, Assignee
601 Wythe Street, Alexandria, VA 22314
Attn: Sarah Balas

PATENT NO. 6,447,683 B1

No. of additional copies

⇨

Burden Hour Statement: This form is estimated to take 1.0 hour to complete. Time will vary depending upon the needs of the individual case. Any comment on the amount of time you are required to complete this form should be sent to the Chief Information Officer, U.S. Patent and Trademark Office, Washington, DC 20231. DO NOT SEND FEES OR COMPLETED FORMS TO THIS ADDRESS. SEND TO: Assistant Commissioner for Patents, Washington, DC 20231.

WASTEWATER UTILITY

Alabama
Montgomery Water Works & Sanitary Sewer Board

Alaska
Anchorage Water & Wastewater Utility

Arizona
Gila Resources

Glendale, City of, Utilities Department

Mesa, City of

Peoria, City of

Phoenix Water Services Department

Pima County Wastewater Management

Arkansas
Little Rock Wastewater Utility

California
Calaveras County Water District

Central Contra Costa Sanitary District

Contra Costa Water District

Crestline Sanitation District

Delta Diablo Sanitation District

Dublin San Ramon Services District

East Bay Dischargers Authority

East Bay Municipal Utility District

El Dorado Irrigation District

Fairfield-Suisun Sewer District

Fresno Department of Public Utilities

Irvine Ranch Water District

Las Virgenes Municipal Water District

Livermore, City of

Lodi, City of

Los Angeles, City of

Los Angeles County, Sanitation Districts of

Napa Sanitation District

Orange County Sanitation District

Palo Alto, City of

Riverside, City of

Sacramento Regional County Sanitation District

San Diego Metropolitan Wastewater Department, City of

San Francisco, City & County of

San Jose, City of

Santa Barbara, City of

Santa Rosa, City of

South Bayside System Authority

South Coast Water District

South Orange County Wastewater Authority

Union Sanitary District

West Valley Sanitation District

Colorado
Boulder, City of

Colorado Springs Utilities

Littleton/Englewood Water Pollution Control Plant

Metro Wastewater Reclamation District, Denver

Connecticut
The Mattabasset District

New Haven, City of, WPCA

District of Columbia
District of Columbia Water & Sewer Authority

Florida
Broward, County of

Fort Lauderdale, City of

Gainesville Regional Utilities

JEA

Miami-Dade Water & Sewer Authority

Orange County Utilities Department

Orlando, City of

Reedy Creek Improvement District

Seminole County Environmental Services

St. Petersburg, City of

Stuart Public Utilities

Tallahassee, City of

Tampa, City of

Toho Water Authority

West Palm Beach, City of

Georgia
Atlanta Department of Watershed Management

Augusta, City of

Clayton County Water Authority

Cobb County Water System

Columbus Water Works

Fulton County

Gwinnett County Department of Public Utilities

Macon Water Authority

Savannah, City of

Hawaii
Honolulu, City and County of

Illinois
American Bottoms Wastewater Treatment Plant

Greater Peoria Sanitary District

Kankakee River Metropolitan Agency

Metropolitan Water Reclamation District of Greater Chicago

North Shore Sanitary District

Wheaton Sanitary District

Indiana
Fort Wayne, City of

Iowa
Ames, City of

Cedar Rapids Wastewater Facility

Des Moines Metro Wastewater Reclamation Authority

Iowa City, City of

Kansas
Johnson County Unified Wastewater Districts

Unified Government of Wyandotte County/Kansas City, City of

Kentucky
Louisville & Jefferson County Metropolitan Sewer District

Louisiana
Sewerage & Water Board of New Orleans

Maine
Bangor, City of

Portland Water District

Maryland
Anne Arundel County Bureau of Utility Operations

Howard County Department of Public Works

Washington Suburban Sanitary Commission

Massachusetts
Boston Water & Sewer Commission

Upper Blackstone Water Pollution Abatement District

Michigan
Ann Arbor, City of

Detroit, City of

Holland Board of Public Works

Lansing, City of

Owosso Mid-Shiawassee County WWTP

Saginaw, City of

Wayne County Department of Environment

Wyoming, City of

Minnesota
Rochester, City of

Western Lake Superior Sanitary District

Missouri
Independence, City of

Kansas City Missouri Water Services Department

Little Blue Valley Sewer District

Nebraska
Lincoln Wastewater System

Nevada
Henderson, City of

New Jersey
Passaic Valley Sewerage Commissioners

New York
New York City Department of Environmental Protection

Rockland County Solid Waste Management Authority/ Sewer District

North Carolina
Cary, Town of

Charlotte/Mecklenburg Utilities

Durham, City of

Metropolitan Sewerage District of Buncombe County

Orange Water & Sewer Authority

Ohio
Akron, City of

Butler County Department of Environmental Services

Columbus, City of

Metropolitan Sewer District of Greater Cincinnati

Northeast Ohio Regional Sewer District

Summit, County of

Oklahoma
Tulsa, City of

Oregon
Clean Water Services

Eugene/Springfield Water Pollution Control

Water Environment Services

Pennsylvania
Philadelphia, City of

University Area Joint Authority, State College

South Carolina
Charleston Commissioners of Public Works

Mount Pleasant Waterworks & Sewer Commission

Spartanburg Sanitary Sewer District

Tennessee
Cleveland, City of

Knoxville Utilities Board

Murfreesboro Water & Sewer Department

Nashville Metro Water Services

Texas
Austin, City of

Dallas Water Utilities

Denton, City of

El Paso Water Utilities

Fort Worth, City of

Gulf Coast Waste Disposal Authority

WERF SUBSCRIBERS

Houston, City of
San Antonio Water System
Trinity River Authority

Utah
Salt Lake City Corporation

Virginia
Alexandria Sanitation Authority
Arlington, County of
Fairfax County Virginia
Hampton Roads Sanitation District
Henrico, County of
Hopewell Regional Wastewater Treatment Facility
Loudoun County Sanitation Authority
Lynchburg Regional WWTP
Prince William County Service Authority
Richmond, City of
Rivanna Water & Sewer Authority

Washington
Edmonds, City of
Everett, City of
King County Department of Natural Resources
Seattle Public Utilities
Sunnyside, Port of
Yakima, City of

Wisconsin
Green Bay Metro Sewerage District
Madison Metropolitan Sewerage District
Milwaukee Metropolitan Sewerage District
Racine, City of
Sheboygan Regional Wastewater Treatment
Wausau Water Works

Australia
South Australian Water Corp.
Sydney Water Corp.
Water Corp. of Western Australia

Canada
Greater Vancouver Regional District
Toronto, City of, Ontario
Winnipeg, City of, Manitoba

Mexico
Servicios de Agua y Drenaje de Monterrey, I.P.D.

New Zealand
Watercare Services Limited

Singapore
Singapore Public Utilities Board

United Kingdom
Yorkshire Water Services Limited

STORMWATER UTILITY

California
Los Angeles, City of, Department of Public Works
Monterey, City of
Santa Rosa, City of

Colorado
Boulder, City of

Georgia
Griffin, City of

Iowa
Cedar Rapids Wastewater Facility
Des Moines Metro Wastewater Reclamation Authority

Kansas
Overland Park, City of

Kentucky
Louisville & Jefferson County Metropolitan Sewer District

Maine
Portland Water District

Minnesota
Western Lake Superior Sanitary District

North Carolina
Cary, Town of
Charlotte, City of, Stormwater Services

Pennsylvania
Philadelphia, City of

Tennessee
Chattanooga Storm Water Management

Washington
Seattle Public Utilities

STATE

Arkansas Department of Environmental Quality
Fresno Metropolitan Flood Control District, Calif.
Urban Drainage & Flood Control District, Colo.

CORPORATE

ADS Environmental Services
The ADVENT Group Inc.
Alan Plummer & Associates
Alden Research Laboratory
Alpine Technology Inc.
Aqua-Aerobic Systems, Inc.
Aquateam–Norwegian Water Technology Centre A/S
BaySaver Inc.
BioVir Laboratories, Inc.
Black & Veatch
Boyle Engineering Corporation
BPR CSO
Brown & Caldwell
Burns & McDonnell
CABE Associates Inc.
The Cadmus Group
Camp Dresser & McKee Inc.
Carollo Engineers Inc.
Carpenter Environmental Associates Inc.
CDS Technologies Inc.
Chemtrac Systems Inc.
CH2M HILL
Damon S. Williams Associates, LLC
David L. Sheridan, P.C.
Earth Tech Inc.
Eco-Matrix
Environmental Engineers International
EMA Inc.
The Eshelman Company Inc.
Finkbeiner, Pettis, & Strout Inc.
Freese & Nichols, Inc.
Frontier Geosciences Inc.
ftn Associates Inc.
Fuss & O'Neill, Inc.
Gannett Fleming Inc.
Golder Associates Inc.
Greeley and Hansen LLC
Hazen & Sawyer, P.C.
HDR Engineering Inc.
HNTB Corporation
HydroQual Inc.
Ingersoll-Rand Energy Systems
Insituform Technologies Inc.
Institute for Environmental Technology & Industry, Korea
Jacobson Helgoth Consultants Inc.
Jason Consultants Inc.
Jordan, Jones, & Goulding Inc.
KCI Technologies Inc.
Kelly & Weaver, P.C.
Kennedy/Jenks Consultants
Komline Sanderson Engineering Corporation
Lawler, Matusky & Skelly Engineers, LLP
Limno-Tech Inc.
Lombardo Associates Inc.
Malcolm Pirnie Inc.
McKim & Creed
MEC Analytical Systems Inc.
Metcalf & Eddy Inc.
MWH
New England Organics
Odor & Corrosion Technology Consultants Inc. (OCTC)
ONDEO Degremont Inc.
Oswald Green, LLC
PA Government Services Inc.
Parametrix Inc.
Parsons
Post, Buckley, Schuh & Jernigan
R&D Engineering/Conestoga Rover & Associates
The RETEC Group
RMC, Inc. (Raines, Melton & Carella)
R.M. Towill Corporation
Ross & Associates Ltd.
Royce Technologies
SAIC Maritime Technical Group
Sear-Brown
Short Elliott Hendrickson, Inc.
Stormwater Management, Inc.
Synagro Technologies, Inc.
Tetra Tech Inc.
Trojan Technologies Inc.
URS Corporation
Veolia Water NATC
Wade-Trim Inc.
Weston Solutions Inc.
Woodard & Curran
WRc/D&B, LLC
WWETCO, LLC
Zoeller Pump Company

INDUSTRY

American Electric Power
ChevronTexaco Energy Research & Technology Company
The Coca-Cola Company
Dow Chemical Company
DuPont Company
Eastman Kodak Company
Merck & Company Inc.
ONDEO Services
Procter & Gamble Company
PSEG Services Corp.
Severn Trent Services Inc.
Shell Global Solutions
Thames Water Plc
United Water Services LLC

Note: List as of 1/31/04

Board of Directors

Research Council

Stormwater Technical Advisory Committee

WERF Product Order Form

As a benefit of joining the Water Environment Research Foundation, subscribers are entitled to receive one complimentary copy of all final reports and other products.Additional copies are available at cost (usually $10).To order your complimentary copy of a report, please write "free" in the unit price column.WERF keeps track of all orders.If the charge differs from what is shown here, we will call to confirm the total before processing.

Name Title

Organization

Address

City State Zip Code Country

Phone Fax Email

Stock #	Product	Quantity	Unit Price	Total
			Postage & Handling	
			VA Residents Add 4.5% Sales Tax	
			Canadian Residents Add 7% GST	
			TOTAL	

Method of Payment: (All orders must be prepaid.)

❑ Check or Money Order Enclosed

❑ Visa ❑ Mastercard ❑ American Express

Account No. Exp. Date

Signature

Shipping & Handling:

Amount of Order	United States	Canada & Mexico	All Others
Up to but not more than:	Add:	Add:	Add:
$20.00	$5.00*	$8.00	50% of amount
30.00	5.50	8.00	40% of amount
40.00	6.00	8.00	
50.00	6.50	14.00	
60.00	7.00	14.00	
80.00	8.00	14.00	
100.00	10.00	21.00	
150.00	12.50	28.00	
200.00	15.00	35.00	
More than $200.00	Add 20% of order	Add 20% of order	

*minimum amount for all orders

Note: Please make checks payable to the Water Environment Research Foundation.

To Order (Subscribers Only):

Log on to www.werf.org and click on the "Product Catalog."

Phone: (703) 684-2470
Fax: (703) 299-0742.

WERF
Attn: Subscriber Services
635 Slaters Lane
Alexandria, VA 22314-1177

To Order (Non-Subscribers):

Non-subscribers may be able to order WERF publications either through WEF (www.wef.org) or IWAP (www.iwapublishing.com).Visit WERF's website at www.werf.org for details.

IWA Publishing's authorised EU representative for General Product Safety Regulations is Diane D'Arras, 15 rue Duret, 75116 Paris, France, e-mail: safety@iwap.co.uk.

Printed and bound by CPI Group (UK) Ltd, Croydon, CR0 4YY
12/05/2026
02108876-0002